maths for advanced chemistry

Mike Robinson with
Mike Taylor

Published in 2002 by:
Nelson Thornes Ltd
Delta Place
27 Bath Road
CHELTENHAM
GL53 7TH
United Kingdom

02 03 04 05 06 / 10 9 8 7 6 5 4 3 2 1

A catalogue record for this book is available from the British Library

ISBN 0 7487 6582 4

Illustrations by Oxford Designers and Illustrators
Page make-up by Mathematical Composition Setters Ltd

Printed and bound in Croatia by Zrinski

Contents

About this book

Chemistry at this level is not a very mathematical subject, although it can seem so if you lack confidence in handling numbers and equations, or interpreting graphs. If you were completely happy with your maths, you would not be reading this. Whether you are held up by a very small point, or some part of the maths seems to be in a foreign language, we can help.

We have listed all the maths topics (from arithmetic to algebra) which your course needs. Each topic has a short chapter. Each chapter has

- a list of objectives for you to achieve
- an introduction to put the topic into context
- an explanation
- some 'Try it Yourself' questions
- the answers to these – but they are at the end of the chapter
- a summary

To help, we have used a simple code.

Try It Yourself These questions are in the text and you should try each one as you come to it. If you can do it, good. If you cannot, it means that you need to think again about the text above the symbol.

KEY FACT *Shaded boxes enclose a key fact of a statement of principle which you should be able to remember and use.*

indicates a section on using a VPAM (visually perfect algebraic method) calculator. These calculators show you the sum you have entered as well as the result.

shows a hint for dealing with the problem. This might be a short cut, a way of remembering, a reminder that your VPAM can do the job.

Exam Questions means a question of the sort you may find in an examination. Try them! Practice makes perfect. We have only given the numerical parts of the answers to these questons.

About the calculator

We have assumed that you will use a VPAM calculator. All our instructions are for the Casio *fx*–83WA which is representative of most. If you have a different calculator, find out from its manual how to do each sum.

About units

We have used SI units throughout, with all the usual conventions for abbreviations and unit names.
Where a name is in the plural and written in full, we have used the plural form of the unit (35 joules, not 35 joule). The plural is more in accord with English speech and grammar and introduces no ambiguity.

Thanks

We would like to thank Beth Hutchins and all the staff at Nelson Thornes for the help and encouragement that they have given to us in preparing this book, and their hard work in transforming our drafts into their final polished form.

Mike Robinson
Mike Taylor

Chapter 1

Introduction

After completing this chapter you should be able to:

- *use your calculator to carry out calculations needed in A-level chemistry*
- *convert very large and very small numbers to* ***standard form*** *(scientific notation)*
- *carry out calculations using standard form*
- *use your calculator to display answers in standard form.*

1.1 Using a calculator

Most calculators suitable for A-level chemistry courses are very similar. There may be slight differences in the way the keys are arranged, but they all come with instruction booklets to help you find out how to do most of the calculations needed.

The instructions which follow are for the Casio Scientific *fx-83WA*. This uses a VPAM operating system, which stands for **Visually Perfect Algebraic Method**.

Fig. 1 Casio fx-83WA calculator.

Most calculations can be keyed in as they are written.

Example

3.1 + 4.2

3 . 1 +
4 . 2 =

The display shows:

Simple calculations involving subtraction, division and multiplication can be handled in the same way as shown above.

Try It Yourself

Exercise 1.1.1

Use your calculator to work out the answers to the following calculations:

1 5.8 ÷ 2.8 **2** 6.4 × 9.7

3 81.9 – 15.7 **4** 4.5 + 3.7 + 13.7

You may have noticed that in question 4 you need only press = to finish the calculation, i.e. that the following sequences of keys both give the same answer.

[4] [.] [5] [+] [3] [.] [7] [=] [+] [1] [3] [.] [7] [=]

[4] [.] [5] [+] [3] [.] [7] [+] [1] [3] [.] [7] [=]

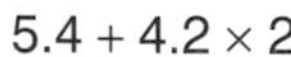

Most calculations can be keyed in just as they are written. There are however some cases where you must be very careful.

5.4 + 4.2 × 2

can be read as

[5] [.] [4] [+]
[4] [.] [2] [=]
[×] [2] [=]

either

(5.4 + 4.2) × 2 which equals 19.2,

The display shows:

then

or

5.4 + (4.2 × 2) which equals 13.8.

The display shows:

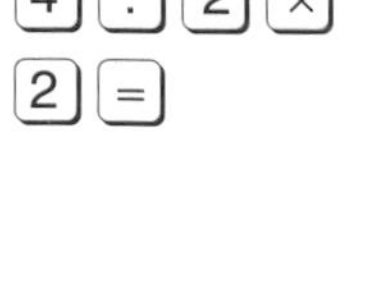

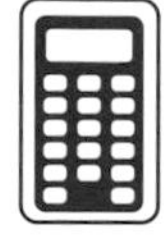

then

KEY FACT *Brackets should always be evaluated first.*

You can work out the values of the expressions above using the brackets available on your calculator. If you want to do it that way, the keys to use for the second calculation are shown below:

[5] [.] [4] [+] [(] [4] [.] [2] [×] [2] [)] [=]

The display shows:

You should decide for yourself whether you find it easier to do calculations with or without brackets.

Occasionally you will have to deal with expressions that are not entered exactly as they are printed.

Example

$$\frac{4.2 \times 3.4}{1.8 \times 2.5}$$

The keys to use are:

The display shows:

The difference between the way the expression is spoken and the way it is entered is that the top line is divided by 1.8 *and* divided by 2.5.

Always:

- *take care when writing expressions to show exactly what you mean*
- *use brackets when necessary to make your meaning clear*
- *take care to enter into your calculator* exactly *what you meant when you wrote the expression*
- *remember that your calculator knows no chemistry! It will do what you tell it to!*

Try It Yourself

Exercise 1.1.2

Use your calculator to work out the following:

1 $4.2 + 3.7 \times 2.4$ **2** $4.2 \times 3.5 + 7.2$ **3** $(4.2 \div 3.5 \times 2.3) + 4$

Your calculator contains many other functions, some of which you will not use in A-level chemistry. Most of them are fairly straightforward. Some are shown below. Try them yourself:

	Example of keys to use	*Display*	*Note*
square root	[√] [3] [.] [4] [=]	1.843908891	(a)
square	[3] [.] [4] [x^2]	11.56	
log	[log] [1] [0] [0] [=]	2.	(b)
ln	[ln] [1] [0] [0] [=]	4.605170186	(c)
$\frac{1}{x}$	[2] [x^{-1}] [=]	0.5	(d)
x^y	[2] [x^y] [3] [=]	8.	

Notes

(a) Just as we say 'root x', the √ *precedes* the number. Other calculators may differ here.

(b) This is needed only for pH calculations in chemistry.

(c) This is unlikely to be needed in A-level chemistry.

(d) Many calculators have a key marked $\frac{1}{x}$ for this function. On the *fx-83WA* it is marked x^{-1}.

The results of the calculations shown above are as they appear on the calculator. Later in the book we will look at how many of those figures should appear in the answer to a calculation, see p. 12.

Try It Yourself

Exercise 1.1.3

Use your calculator to find these:

1 √9.32 **2** 3.4^2 **3** log 17 **4** ln 17

1.2 Scientific notation – a different way of expressing numbers

Scientists often have to handle very large or very small numbers. The number of atoms in 12.000 g of carbon is 602 252 000 000 000 000 000 000. The distance between the centres of the two carbon atoms in an ethane molecule is about 0.000 000 015 cm. These numbers are not only difficult to appreciate, it is very easy to misread them and enter them incorrectly into a calculator. It also makes estimating answers to calculations in which these (or similar) numbers appear, very difficult.

A simpler way to deal with such numbers is to express them in terms of **powers of 10**.

Very large numbers

$$
\begin{aligned}
9381 &= 938.1 \times 10 \\
&= 93.81 \times 10 \times 10 \\
&= 9.381 \times 10 \times 10 \times 10 \\
&= 9.381 \times 10^3 \quad (10 \times 10 \times 10 = 10^3)
\end{aligned}
$$

9.381×10^3 is referred to as **standard form** (or scientific notation). Any number can be written in this way; it is always expressed as a number between 1 and 9.9 (or 9.99999999 if using a 10 figure calculator), multiplied by 10 to the appropriate power.

Example

$$\begin{aligned} 42\,396.75 &= 4239.675 \times 10 \\ &= 423.9675 \times 10 \times 10 \\ &= 42.396\,75 \times 10 \times 10 \times 10 \\ &= 4.239\,675 \times 10 \times 10 \times 10 \times 10 \\ &= 4.239\,675 \times 10^4 \quad (10 \times 10 \times 10 \times 10 = 10^4) \end{aligned}$$

Try it Yourself

Exercise 1.2.1

Write the following numbers in standard form:

1 101 325 (normal atmospheric pressure in Pa)

2 299 700 000 (speed of light in a vacuum in m s^{-1})

3 6 371 000 (radius of the Earth in m)

4 13 600 (density of mercury in kg m^{-3})

5 176 000 000 000 (charge : mass ratio of an electron in C kg^{-1})

Very small numbers

So far we have dealt with very large numbers. Scientific notation can also handle very small numbers:

$$\begin{aligned} 0.0015 &= \frac{0.015}{10} \\ &= \frac{0.15}{10 \times 10} \\ &= \frac{1.5}{10 \times 10 \times 10} \\ &= \frac{1.5}{10^3} \\ &= 1.5 \times 10^{-3} \end{aligned}$$

KEY FACT $\frac{1}{10^3} = 10^{-3}$, *moving the 10 from the bottom line to the top involves changing the sign.*

Try It Yourself

Exercise 1.2.2

Write each of the following numbers in standard form:

1 0.000 000 000 000 000 000 160 (charge on an electron in C, coulombs)

2 0.000 000 015 (distance between two carbon atoms in an ethane molecule in cm)

3 0.000 000 000 005 29 (radius of a hydrogen atom in m)

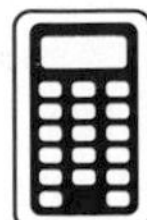

Entering numbers in standard form into your calculator

Powers of 10 are entered in a calculator using the key marked either [EXP] or [EE].

For example, 3.14×10^3 would be entered as:

[3] [.] [1] [4] [EXP] [3]

To enter a negative power of 10, e.g. 4.27×10^{-8}:

[4] [.] [2] [7] [EXP] [–] [8]

or

[4] [.] [2] [7] [EXP] [(–)] [8]

HINT

The above sequence may differ slightly on different calculators.

Calculating using standard notation

KEY FACT *When multiplying powers of 10, the powers are* ***added.***

Thus

$$10^x \times 10^y = 10^{(x+y)}$$

KEY FACT *When dividing powers of 10, the powers are* ***subtracted.***

Thus

$$\frac{10^x}{10^y} = 10^{(x-y)}$$

When dividing, the order in which the powers are subtracted is important!

Examples

1

$$\begin{aligned} 4200 \times 3800 &= 4.2 \times 10^3 \times 3.8 \times 10^3 \\ &= 4.2 \times 3.8 \times 10^6 \\ &= 1.596 \times 10^7 \end{aligned}$$

2

$$\begin{aligned} \frac{4200}{3800} &= \frac{4.2 \times 10^3}{3.8 \times 10^3} \\ &= \frac{4.2}{3.8} \\ &= 1.105 \end{aligned}$$

Controlling the calculator display

Some calculators will display the answer in the form in which you enter it. For example, the first example shown above can be entered into the calculator either in the form shown on the left-hand side of the equals sign:

[4] [2] [0] [0] [×] [3] [8] [0] [0] [=]

or in the form shown on the right-hand side:

[4] [.] [2] [EXP] [3] [×] [3] [.] [8] [EXP] [3] [=]

In the first case, the result will be displayed by many calculators as 15 960 000. In the second case, it may be displayed as 1.596^{07} (which means 1.596×10^7).

Such calculators switch from one type of display to the other when they have to. So a calculator which has displayed an answer of 15 960 000 will switch to scientific notation only when the answer requires more zeros than it can display.

However, the answer displayed by the Casio *fx-83WA* will depend on the **mode** to which it is set. Enter the following on your calculator:

[MODE] [MODE] [MODE]

This causes the calculator to show the choices for the way in which it displays the answers to calculations. You should see:

If you choose 3 (press key [3]) followed by [1], you ensure that the calculator expresses the values of all calculations in non-scientific notation (unless it is forced into scientific notation because the answer contains too many zeros) regardless of the form in which they are entered.

If you choose [2] (press key [2])the calculator displays all answers in scientific notation, again regardless of the form in which they are entered. After you have pressed [2] you have a further choice to make, which is the number of figures (or **significant figures**, see p. 12) that appear in the display. Choosing [4] results in a display containing four digits, i.e. three decimal places.

To return the Casio *fx-83WA* calculator to displaying answers in non-scientific notation, you need to select option 3 from the above menu. The sequence of keys for this is shown below.

[MODE] [MODE] [MODE] [3]

Then press [1].

Try It Yourself

Exercise 1.2.3

1 You should experiment with your calculator, consulting the instruction leaflet where necessary, to make sure that you can switch from one type of display to another quickly and confidently. Try entering each of the calculations below as they are printed so that you can obtain the answer in the form you want it.

(a) $5362 \times 429.4 = 2\,302\,442.8$

(b) $5.36 \times 10^3 \times 4.294 \times 10^2 = 2.301\,584 \times 10^6$ (displayed as $2.301\,584^{06}$)

Notice how your calculator displays the answer to the first example above.

2 Convert the numbers in the following expressions to standard form and work them out:

(a) 345.2×76.42 (b) $0.0376 \times 0.004\,27$ (c) $\dfrac{456.7}{0.007\,23}$

Summary

You should now be able to:

- use your calculator to show that, e.g.

$$\frac{3.72 \times 4.96}{2.64 \times 1.08} = 6.471\,380\,471$$

- convert, e.g., 196 000 and 0.0196 to **standard form** as 1.96×10^5 and 1.96×10^{-2} respectively
- use your calculator to show that, e.g., $1.46 \times 10^3 \times 4.96 \times 10^{-5} = 7.24 \times 10^{-2}$
- make your calculator display the above results as 7.24×10^{-2} or 0.0724.

Answers to Try It Yourself Questions

The number of figures displayed in your answer will depend on the calculator used and the mode in which it is working. These answers are obtained using a Casio *fx-83WA* calculator.

Exercise 1.1.1, *p. 2*
1 2.071 428 571 **2** 62.08 **3** 66.2 **4** 21.9

Exercise 1.1.2, *p. 4*
1 13.08 **2** 21.9 **3** 6.76

Exercise 1.1.3, *p. 5*
1 3.052 867 504 **2** 11.56
3 1.230 448 921 **4** 2.833 213 344

Exercise 1.2.1, *p. 6*
1 $1.013\,25 \times 10^5$ Pa **2** $2.997\,000\,00 \times 10^8$ m s^{-1}
3 $3.700\,00 \times 10^5$ m **4** 3.60×10^2 kg m^{-3}
5 $1.760\,000\,000\,00 \times 10^{11}$ C kg^{-1}

Exercise 1.2.2, *p. 6*
1 1.60×10^{-20} C **2** 1.5×10^{-8} cm
3 5.29×10^{-12} m

Exercise 1.2.3, *p. 9*
2 (a) $2.638\,018\,40 \times 10^4$ (b) $1.605\,52 \times 10^{-4}$
(c) $6.316\,735\,823 \times 10^4$

Chapter 2

Handling numbers

After completing this chapter you should:

- *understand the meaning of the term* ***significant figures***
- *be able to give answers to an appropriate number of significant figures*
- *be able to estimate the answer to a calculation quickly and reliably.*

2.1 Uncertainty in observations

A lot of work done by chemists involves **qualitative observations** (those not involving numbers). Examples include:

- many reducing agents turn acidified potassium dichromate(VI) *from orange to green*
- increasing the temperature *increases rate of reaction*
- sodium salts *produce a yellow colour in a flame test.*

However, it is often necessary to make measurements which are then used in calculations, i.e. to make **quantitative observations**. Examples include:

- dissolving *40 g* of sodium hydroxide to make *250 cm^3* of solution gives a concentration of 160 g dm^{-3}
- *doubling* the concentration of a reagent in a particular reaction *doubles* the rate.

It is important when you make measurements to understand how reliable they are, as it will affect the reliability of your answer. Obviously if your initial measurements are approximate, your final answer will be only approximate; if your initial measurements are more accurate, your final answer will be more accurate.

Imagine that you have prepared a solution of sodium carbonate and that you need to know its concentration. You will have weighed out the sodium carbonate, and then dissolved it to make a measured volume of solution. Most balances in chemical laboratories read to the nearest 0.01 g or better. You may have weighed out *15.87 g* of sodium carbonate.

Your sample may contain as little as 15.865 g or as much as 15.874 g; any mass between these two figures would register as 15.87 g on a balance reading to 0.01 g. 15.864 g would have registered as 15.86 g, and 15.875 g would have registered as 15.88 g. (See *Fig. 1.*)

Imagine that you dissolved the sodium carbonate to make 250 cm^3 of solution. You might have done this by adding the solid to 250 cm^3 of **water** measured using a measuring cylinder (you will probably learn that this is a very poor way of preparing a solution of *accurately* known concentration), or you might have used a graduated (volumetric) flask to prepare 250 cm^3 of **solution**. In the first case you will have used somewhere between 248 and 252 cm^3 of water, in the second case you will have prepared between 249.5 and 250.5 cm^3 of solution.

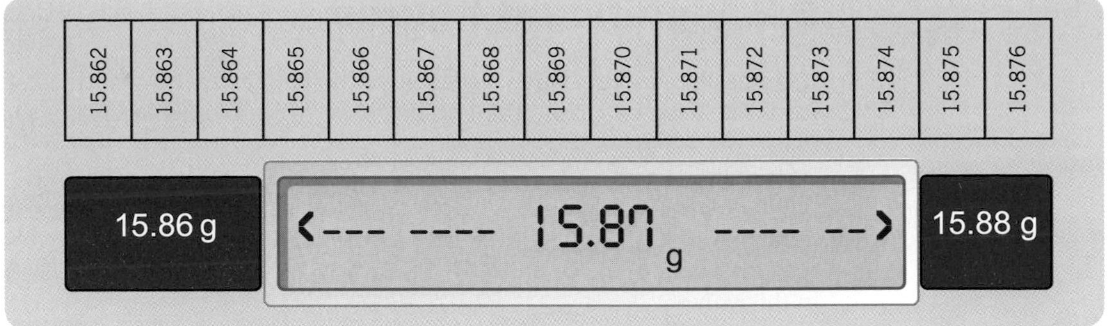

Fig. 1 *A balance reading of 15.87 g.*

KEY FACT *The inaccuracies/uncertainties in the values you have measured mean that there will be some uncertainty over the concentration of your final solution.*

You could calculate the concentration of the solution as follows.

Example

15.87 g of sodium carbonate in 250 cm^3

$\frac{15.87}{250}$ g of sodium carbonate in 1 cm^3

$$\frac{15.87 \times 1000}{250} = 63.48 \text{ g of sodium carbonate in 1 dm}^3 \text{ (1000 cm}^3\text{)}$$

The concentration of the solution is thus 63.48 g dm^{-3}.

However, using a measuring cylinder for the water, you could have had as much as 15.874 g dissolved in as little as 248 cm^3 of solution, giving a concentration of:

$$\frac{15.874 \times 1000}{248} = 64.008\ 064\ 52 \text{ g dm}^{-3}$$

The answer is the one shown on my calculator!

This value, using the *largest* mass (top line of equation) in the *smallest* volume (bottom line of equation) is the *maximum* possible value of the concentration.

On the other hand, you could have had as little as 15.865 g dissolved in as much as 252 cm^3 of solution, giving a concentration of:

$$\frac{15.865 \times 1000}{248} = 62.956\ 349\ 21 \text{ g dm}^{-3}$$

This value, using the *smallest* mass (top line of equation) in the *largest* volume (bottom line of equation) is the *minimum* possible value of the concentration.

So your value of 63.48 g dm^{-3} could really be anywhere between 64.008 064 52 g dm^{-3} and 62.956 349 21 g dm^{-3}.

2.2 Significant figures

It is possible to carry out a rigorous mathematical analysis of this particular problem, but fortunately we can avoid that! To do so we need to use the idea of **significant figures** (s.f.)

The balance reading of 15.87 g contains four figures, and is said to be to four significant figures. Think of significant figures as those we know reliably; we do not know whether we really have 15.870 g or 15.871 g, or even 15.874 g, i.e. we do *not know* what the fifth figure is in this value.

The number of significant figures in a value is those that we know *excluding any zeros at the beginning*.

Try It Yourself

Exercise 2.2.1

How many significant figures are there in the following numbers?

1 1.075 **2** 1.0752 **3** 0.1075 **4** 0.000 107 50

The volume of solution prepared using a measuring cylinder could be between 248 and 252 cm^3. This makes the final digit unreliable; we know only the first two digits reliably, so that we know the volume in this case to only two significant figures.

If our solution contains 15.87 g in 250 cm^3, it contains 4 times this mass in 1 dm^3, i.e. 63.48 g. However, to claim that the concentration is 63.48 g dm^{-3} is not justified, given the uncertainties in the values we used.

KEY FACT *A useful rule of thumb is that the number of significant figures in the answer cannot be greater than the* smallest number of **significant figures** *in the measurements used. In this case a mass to four significant figures and a volume to only two significant figures results in a concentration* known reliably *to only two significant figures – probably most conveniently written as 63 g dm^{-3}.*

Remember that just because your calculator produces a lot of figures, there is no guarantee that they are reliable (or *significant*). My calculator produced answers to *ten* significant figures in the calculations above; only the first *two* are reliable!

If however we had prepared the solution using a graduated flask, we would have known the volume to three significant figures. The concentration would then have been reliable to three significant figures, and may be quoted as 63.5 g dm^{-3}.

Try It Yourself

Exercise 2.2.2

1. What would be the concentration of solution if the measurements had been made using a balance reading to 0.1 g and a volumetric flask?
2. What would be the concentration of solution if the measurements had been made using a balance reading to 0.001 g and a volumetric flask?

Most chemical calculations involve multiplication and/or division. Under these circumstances the 'rule' already given, that the number of significant figures in an answer cannot be greater than those in the measurement, is correct.

2.3 Significant figures in addition or subtraction

However, when a calculation involves addition or subtraction, for example when calculating a temperature rise from temperatures taken at the start and end of a chemical reaction, a different rule applies.

KEY FACT *In this case, the rule is that the number of* **decimal places (d.p.)** *in the answer should be the same as those in the readings. For example, if the initial and final temperatures in a chemical reaction are measured as 15 °C and 25 °C respectively, the temperature rise is 10 °C (not 10.0 °C) – the readings and temperature rise are all to zero decimal place.*

If the readings were taken with a more sensitive thermometer and found to be 15.0 °C and 25.0 °C, then it is appropriate to quote the temperature rise as 10.0 °C – the readings and the temperature rise are all to one decimal place.

2.4 Making estimates

It is often useful to be able to estimate the answer to a calculation. You may need to have only an approximate answer, which you can obtain without using a calculator. Alternatively, you might find it a useful way of checking that an answer you have obtained with your calculator is about right – that you have not omitted a few zeros, or entered a decimal point in the wrong place.

You should always aim to estimate an answer without using a calculator. That means that you need to simplify the calculation to one that you can do in your head. You will therefore need to round off, or approximate each of the numbers involved. The more you approximate, the less accurate your estimate will be. You should therefore approximate as little as possible, but still reduce the figures to ones you can handle in your head.

HINT *You should remember that there is no set way of arriving at an estimate. Each calculation must be approached slightly differently, with a view to making the best estimate that you can.*

Example

Method 1 $6.57 \times 2.43 \simeq 7 \times 2$ *Round off each figure to the nearest whole number.*

$= 14$

Method 2 $6.57 \times 2.43 \simeq 6.5 \times 2.5$ *Round to the nearest half.*

$= \frac{13}{2} \times \frac{5}{2}$ *Express as fractions.*

$= \frac{65}{4}$ *Multiply out fractions.*

$\simeq 16$ *Simplify fraction.*

Accurate answer 15.9651

Question: Bearing in mind what has already been said about significant figures (see p. 12), what should the accurate answer really be, starting from 6.57 and 2.43?

Answer: the original data is given to three significant figures; the answer can then be given to only three significant figures, in which case it should be 16.0.

You will see that Method 1, which is very approximate, gives an answer some way from the accurately calculated value; whilst Method 2, which takes a little longer, gives an answer much closer to the correct one. Notice, though, that both methods show the decimal point is in the correct place.

Example

Estimate 432×384.

Method 1 $432 \times 384 \simeq 400 \times 400$ *Round off each figure to the nearest hundred.*

$= 160\,000$

Method 2 $432 \times 384 \simeq 450 \times 400$ *Round to the nearest fifty.*

$= 180\,000$

Accurate answer 165 888

Question: How should the accurate answer be displayed, allowing for the number of significant figures in the original data?

Answer: the original data is given to three significant figures; the answer can then be given to only three significant figures, in which case it should be 166 000. However, this way of presenting it does not make clear the number of significant figures being quoted. In this case you should use scientific notation (see p. 5) writing it as 1.66×10^5.

You will see that in this case Method 1, which involves rounding only to the nearest hundred, gives an estimate closer to the correct answer than Method 2, which, on face of it, appears more accurate.

Question: Have a look at the figures involved, and see if you can see why Method 1 gives the closer answer.

Answer: In method 1, rounding to the nearest hundred reduces 432 to 400 and increases 384 to 400. In Method 2, rounding to the nearest fifty increases both values. The two approximations in Method 1, to some extent, compensate for each other (or cancel). It is not always easy to spot in advance which method will be more reliable!

Estimating in standard form

Standard form makes estimating answers more straightforward.

Example

Estimate 4325 × 3841.

4325×3841	$= 4.325 \times 10^3 \times 3.841 \times 10^3$	*Write each number in standard form.*
	$= 4.325 \times 3.841 \times 10^6$	
	$\simeq 4 \times 4 \times 19^6$	*Round off each number.*
	$= 16 \times 10^6$	*Evaluate.*
	$= 1.6 \times 10^7$	*Write in standard form.*
Accurate answer	$1.661\ 232\ 5 \times 10^7$	

As each of the original numbers is given to only four significant figures, the accurate answer should be given as 1.661×10^7.

Try It Yourself

Exercise 2.4.1

For each of the following, estimate the answer without using your calculator. You may wish to do this by more than one method. Finally, use your calculator to obtain an accurate answer, and compare it with your estimate.

1. 35×27 **2** 2506×4207 **3** $\frac{6346}{2519}$

Summary

You should now be able to:

- state the number of significant figures in a number, e.g. 1.2479 (5 s.f.)
- be able to write a number to a specified number of significant figures, e.g. 123.459 to four significant figures (123.5)
- be able to estimate the result of a calculation without using a calculator, e.g. $17.5 \times 4.8 \simeq 90$.

Answers to Try It Yourself Questions

Exercise 2.2.1, *p. 12*
1 4 **2** 5 **3** 4 **4** 5

Exercise 2.2.2, *p. 12*
1 63.6 g dm^3 **2** 63.6 g dm^{-3}

Exercise 2.4.1, *p. 15*
1 1050 (calculator gives 945)
2 11 000 000 (calculator gives 10 542 742)
3 2.5 (calculator gives 2.519 253 672 – but since the data is to only 4 s.f., the answer should be given as 2.519)

Chapter 3

Equations

After completing this chapter you should:

- *appreciate what we can learn from an equation*
- *be able to substitute values into an equation*
- *know that it is important to use the correct units when carrying out calculations using equations.*

3.1 What is an equation?

An **equation** shows the relationship between two or more **variables**. Examples you may already have met include the following.

The area of a circle is always $\pi \times$ the square of the radius of the circle:

In equation form $A = \pi r^2$

The volume of a sphere is always $\frac{4}{3} \times \pi \times$ the cube of its radius:

In equation form $V = \frac{4}{3}\pi r^3$

The number of moles of a compound is always given by the mass of the compound divided by its relative molecular mass:

In equation form $n = \dfrac{m}{M_r}$

The concentration of a solution is always given by the number of moles of solute divided by the volume of the solution:

In equation form $c = \dfrac{n}{V}$

3.2 Using equations – substituting numbers

Quite often using equations simply involves putting in the numbers and working them out (using your calculator if appropriate).

There is however one important check to make – you need to take care over *units*. For example, the concentration of a solution, c, is given by the equation below:

$$c = \frac{n}{V}$$

where n = number of moles, V = volume of solution.

If n is measured in moles and V in decimetres cubed (dm^3), then concentration will be calculated in moles per decimetre cubed ($mol\ dm^{-3}$). The point to remember is that the units of c depend on those used for n and V.

KEY FACT *The units of a quantity calculated from an equation depend on the units of the data you used.*

In the next chapter we will discuss in more detail which units are appropriate in calculations (see p. 23).

Example

If 0.0125 mol are dissolved in 0.250 dm^3, the concentration will be:

$$\frac{0.0125}{0.250} = 0.0500\ \text{mol dm}^{-3}$$

The initial data are to three significant figures, so the answer should be given to three significant figures.

HINT *Your calculator may display the answer in a number of different forms, including:*

0.05   0.050 000 00 5^{-02} $5.000\ 000\ 00^{-02}$

Which one you see depends on your particular calculator, and possibly on the **mode** *you have chosen. If you are uncertain about this, have another look at p. 8*

Try It Yourself — Exercise 3.2.1

1 Under appropriate conditions, the pressure (P), volume (V), temperature (T) and number of moles (n) of a gas are related by the following equation:

$$V = \frac{nRT}{P}$$

(a) If n, R, T, and P are known, what can be found from the equation?

(b) Calculate the volume which 1 mol of argon will occupy at a temperature of 350 K and a pressure of 1.00×10^5 Pa, if the value of the Gas Constant, R, is 8.31 J mol^{-1} K^{-1}.

2 The heat needed (Q) to raise the temperature of mass, m, of water by ΔT is given by the following equation:

$$Q = ms\Delta T$$

The value of s (the specific heat capacity) is 4200 J kg^{-1} K^{-1}.

Calculate the heat required to raise the temperature of 1.25 kg of water by 8.00 K.

3 If Q kJ of energy are released when m g of a substance is burned, the molar enthalpy of combustion (in kJ mol^{-1}), ΔH_c, is given by the following equation:

$$\Delta H_c = -\frac{Q \times M_r}{m}$$

where M_r is the relative molecular mass of the substance involved.

HINT *The minus sign is included in the equation as heat evolved is considered negative. (You should read the chapter on enthalpy in your normal text book if you did not know this.)*

When 1.00 g (0.00100 kg) of methane (CH_4) is burned, 55.7 kJ of energy are released. Calculate the molar enthalpy of combustion of methane.

Summary

You should now:

- understand that an equation shows the *constant* relationship between two or more variables
- be able to use an equation e.g. concentration $= \frac{\text{number of moles}}{\text{volume}}$, to calculate e.g. concentration, given number of moles and volume
- know that when calculating a value using an equation, its units will depend on the units of the data from which it is derived.

Answers to Try It Yourself Questions

Exercise 3.2.1, *p. 17*

1 (a) The volume, V, by substituting the values of n, R, T and P into the equation.
(b) 0.0291 m^3.

2 42 000 J

3 −891 kJ mol^{-1}

Chapter 4

Rearranging equations

After completing this chapter you should be able to:

- *rearrange an equation so that one quantity is on one side and the others are on the other side*
- *appreciate the importance of using* ***consistent units***
- *be able to work out the* ***units*** *for a value calculated using an equation.*

4.1 Introduction to equations

Equations are easy to handle where the quantity you want to calculate appears on the left-hand side of the equation, and the values from which it is to be calculated all appear on the right-hand side.

For example, it is easy to use the equation $n = \frac{m}{M_r}$ to calculate n when you know the mass and the relative molecular mass. It is more difficult to calculate M_r which appears on the right-hand side of the equation, mixed up with the other quantities.

Subject of an equation

The single quantity on the left-hand side is the **subject of the equation**.

It is therefore useful to be able to rearrange an equation to change its subject.

4.2 Rules for rearranging equations (or changing the subject)

They can be summarised in a single rule as follows:

RULE *Always do the same to both sides of the equation.*

This means that if you add something to the left-hand side of an equation, you must add the same thing to the right-hand side.

So if

$$a = b - c$$
$$a + c = b - c + c$$
$$a + c = b$$

RULE *If you subtract something from the left-hand side, you must subtract the same thing from the right-hand side.*

So if

$$d = e + f$$
$$d - f = e + f - f$$
$$d - f = e$$

We could describe what has happened here as 'taking f to the other side and changing its sign', so that $+f$ on the right-hand side becomes $-f$ on the left-hand side.

RULE *If you divide the left-hand side by something you must divide the right-hand side by the same thing.*

So if

$$g = hf$$
$$\frac{g}{f} = \frac{hf}{f}$$
$$\frac{g}{f} = h \text{ (which is the same as } h = \frac{g}{f}\text{)}$$

RULE *If you multiply the left-hand side by something, you must multiply the right-hand side by the same thing.*

So if

$$j = \frac{k}{m}$$
$$j \times m = \frac{k \times m}{m} = k$$

HINT *You must be particularly careful when using the last two methods to make sure that when you divide or multiply you apply it to the* whole *of each side of the equation.*

so if

$$\frac{n}{p} = q + r$$
$$n = p \times (q + r)$$

Example

1 To use the equation $n = \frac{m}{M_r}$ to calculate the mass (m) of substance which contains a particular number (n) of moles, we need to make m the subject of the equation:

$$n = \frac{m}{M_r} \quad \text{original equation}$$

$$nM_r = \frac{mM_r}{M_r} \quad \text{multiply both sides by } M_r$$

$$nM_r = m \quad \text{cancel}$$

which is the same as $m = nM_r$

Have a look at what we have done here. We wanted to make m the subject of the equation, that is to have it alone on the left-hand side. In the original equation it is not alone, but with M_r. We have separated m from M_r.

2 To calculate the relative molecular mass (M_r) of a substance from the mass (m) of a given number (n) of moles, we need to make M_r the subject of the equation:

$$n = \frac{m}{M_r} \quad \text{original equation}$$

$$nM_r = \frac{mM_r}{M_r} \quad \text{multiply both sides by } M_r$$

$$nM_r = m \quad \text{cancel}$$

$$M_r = \frac{m}{n} \quad \text{divide both sides by } n$$

What have we done here? We wanted to make M_r the subject of the equation. In the original equation it was associated with m. We have separated M_r from m.

Changing the subject of the equation is a useful skill. It reduces the number of equations that you have to memorise.

Chemists often have to:

- calculate the number of moles in a given mass of substance
- calculate the mass of substance which contains a particular number of moles
- calculate the relative molecular mass from the mass of a given number of moles.

All three calculations use the relationship:

$$n = \frac{m}{M_r}$$

but in a different form, or with different quantities as the subject. If you can rearrange this equation to make m or M_r the subject, then you need to memorise only the equation above.

If not, you will have to memorise also:

$$m = n \times M_r \quad \text{and} \quad M_r = \frac{m}{n}$$

which are the equations we have just obtained.

Try It Yourself

Exercise 4.2.1

1 The pressure (P), volume (V) and temperature (T) of an ideal gas are related by the equation given below:

$PV = nRT$ where n = the number of moles of gas
R = the Ideal Gas Constant

Make P the subject of this equation.

2 Make n the subject of the equation in question 1.

3 Propanone (CH_3COCH_3) reacts with iodine in acidic solution to give iodopropanone and hydrogen iodide as shown in the following equation:

$$CH_3COCH_3 + I_2 \rightarrow CH_3COCH_2I + HI$$

The relationship between the rate of this reaction, r, and the concentration of the reactants is given by the following equation:

$r = k[CH_3COCH_3][H^+]$ where k = the rate constant and
[X] represents the concentration of species X.

Make k the subject of this equation.

4 The relationship between the change in free energy (ΔG), the enthalpy change (ΔH), the entropy change (ΔS) and the temperature (T) is shown in the equation below:

$$\Delta G = \Delta H - T\Delta S$$

Make ΔS the subject of this equation.

5 The relationship between the concentration of hydrogen ions, of ethanoate ions and ethanoic acid molecules in an aqueous solution of ethanoic acid is given by the equation below:

$$K_a = \frac{[H^+][CH_3COO^-]}{[CH_3COOH]}$$

where K_a = acid dissociation constant of ethanoic acid.
Make $[H^+]$ the subject of this equation.

Squares

RULE *Sometimes the quantity which is to be the subject of an equation appears in squared form. To deal with this we take the square root of both sides of the equation.*

Example

In an aqueous solution of pure ethanoic acid, the concentrations of hydrogen and ethanoate ions are equal. The equation given in question 5 above becomes

$$K_a = \frac{[H^+]^2}{[CH_3COOH]}$$

To make $[H^+]$ the subject of this expression:

$K_a = \frac{[H^+]^2}{[CH_3COOH]}$	original equation
$K_a[CH_3COOH] = [H^+]^2$	multiply both sides by $[CH_3COOH]$
$[H^+] = \sqrt{K_a[CH_3COOH]}$	take the square root of both sides

Try it Yourself

Exercise 4.2.2

1 If a mixture of ethanoic acid and ethanol is heated under reflux with some concentrated sulphuric acid as a catalyst, an equilibrium is established between these compounds and the products, ethyl ethanoate and water. The equation for the reaction is:

$$CH_3COOH + C_2H_5OH \rightleftharpoons CH_3COOC_2H_5 + H_2O$$

The concentrations of each of the species at equilibrium are controlled by the equilibrium law as shown below:

$$K_c = \frac{[CH_3COOC_2H_5][H_2O]}{[CH_3COOH][C_2H_5OH]}$$

HINT *Remember [X] represents the concentration of species X and K_c is the equilibrium constant.*

If a mixture of ethanoic acid and ethanol, each of concentration 1 mol dm^{-3}, is brought to equilibrium, it can be shown that the concentration of ethyl ethanoate, x, will be given by the following equation:

$$K_c = \frac{x^2}{(1-x)^2}$$

Make x the subject of this equation.

HINT *If you want to know where the above equation has come from, it is shown in the Appendix at the end of this chapter, p. 29.*

4.3 Units

The distance from Aberystwyth to Edinburgh is 335 miles. It is also 536 km. The fact that the numbers are different has not altered the distance between the two places. In the same way, a bag of sugar weighs 2.2 lb or 1 kg – the different numbers do not alter the mass of sugar.

We are now going to carry out the same calculation twice using *different* units to see the effect.

Try It Yourself

Exercise 4.3.1

1 (a) Under appropriate conditions, the pressure (P), volume (V), temperature (T) and number of moles (n) of a gas are related by the following equation:

$$pV = nRT$$

Rearrange this to make R the subject of the expression.

You should obtain:

$$R = \frac{pV}{nT}$$

HINT *If you cannot obtain this, have another look at pp. 19–21.*

(b) In an experiment, it was found that 1.00 mol of argon occupied 0.0244 m^3 at 1.01×10^5 Pa pressure and 298 K.

Use these values to calculate the value of the Gas Constant, R.

HINT *You should obtain a value of 8.27 – notice that the data was all given to three significant figures, so this is the appropriate number to quote in the answer (see pp. 12–13).*

2 In an experiment carried out by different scientists, it was found that 1 mol of argon occupied 24.4 l at 1 atm and 298 K.
Use these values to calculate the value of the Gas Constant.

HINT *You should obtain a value of 0.0819 – again to three significant figures.*

Notice that the values of R in the two questions above are different! They are both correct – they are simply measured in different units.

The value of R obtained from the first experiment (8.27) is obtained measuring pressure in pascals, temperature in kelvins and volume in metres cubed (*see Table 1*).

Table 1 *The Gas Constant*

R	Pressure in:	Temperature in:	Volume in:
8.27	Pa	K	m^3
0.0819	atm	K	l

KEY FACT *Whenever the value R = 8.27 is used in the Gas Equation, the pressure, temperature and volume* ***must be measured in pascals, kelvins and metres cubed****.*

If you use the value 0.0819 for the Gas Constant, then pressure must be measured in atmospheres, temperature in kelvins and volume in litres.

There are thus 'correct' combinations of units to use in this equation (Pa/K/m^3 or atm/K/l). These are often described as a **consistent set of units**.

The two different values of *R* obtained have different units. It is useful (almost essential) to be able to work out what the units are. In many calculations, the units are obvious; in others, such as this, they are far from obvious.

Working out units from measurements

When you calculated the value for *R* in question 1(b), you inserted numbers into an equation:

$$R = \frac{pV}{nT} = \frac{1.01 \times 10^5 \times 0.0244}{1.00 \times 298} = 8.27$$

Rewriting this, including the units of each quantity, gives:

$$R = \frac{1.01 \times 10^5 (\text{Pa}) \times 0.0244\ (\text{m}^3)}{1.00\ (\text{mol}) \times 298\ (\text{K})}$$

Rewriting this with *only* the units gives:

$$R = \frac{\text{Pa} \times \text{m}^3}{\text{mol} \times \text{K}}$$

The units can now be 'multiplied out' as if they were just ordinary numbers.

Pascals are pressure units representing the force measured in newtons on an area measured in square metres. The top line is thus:

$$\text{Pa} \times \text{m}^3 = \text{N m}^{-2} \times \text{m}^3 = \text{N m}$$

HINT $m^{-2} \times m^3 = m$, *just as* $a^{-2} \times a^3 = a^{-1}$, *powers added when multiplying. Newton-metres (N m) are the same as joules (J).*

The units can now be simplified to:

$$\frac{\text{N m}}{\text{mol} \times \text{K}} = \frac{\text{J}}{\text{mol K}} \text{ which we write as J mol}^{-1}\,\text{K}^{-1}.$$

Thus the value of the Gas Constant is 8.27 J mol^{-1} K^{-1}.

Try it Yourself **Exercise 4.3.2**

1 Try repeating the process above for the calculation using atmospheres, litres and kelvins (from question 2, p. 24).

HINT

You should find that the units of R are now atm l mol^{-1} K^{-1}.

Commonly used units – SI units

Reference has already been made to the need to use *consistent sets of units*. There are several such sets, but the one most commonly used by scientists is the *Système International d'Unités* **(SI)**. In this:

KEY FACT

- *The unit of length is the metre (m).*
- *The unit of mass is the kilogram (kg).*
- *The unit of time is the second (s).*

Other units are said to be *derived* from these; some examples are shown below.

KEY FACT

- *Volume is measured in metres cubed (m^3).*
- *Pressure is measured in pascals (Pa (formerly N m^{-2})).*
- *Density is measured in kilograms per metre cubed (kg m^{-3}).*

In some cases these units result in very large or very small numbers; it is then acceptable to use more convenient units, but great care needs to be taken when using them in equations.

KEY FACT

Examples of more convenient units used by chemists include:

- *grams, for measuring masses on a laboratory scale*
- *centimetres cubed or decimetres cubed (cm^3 or dm^3), for measuring volumes of laboratory scale reagents*
- *grams per centimetre cubed (g cm^{-3}) for measuring density.*

Rules for writing units

You will see above that only certain units have capital letters. This is restricted to units named after an individual scientist. Thus pressure is measured in Pa (named after Pascal), force in N (named after Newton), energy in J (named after Joule). However, when the units are written in full (e.g. pascal) it is normal to do so *without* a capital letter.

Some units involve a *combination* of units (**compound units**). Examples include:

- density measured in kilograms per metre cubed (kg m^{-3})
- enthalpy measured in joules per mole (J mol^{-1})
- concentration measured in moles per decimetre cubed (mol dm^{-3}).

Notice that in these there is a space between each of the components (mol dm^{-3}, *not* $moldm^{-3}$).

Try it Yourself

Exercise 4.3.3

Note Concentrations are usually measured in mol dm^{-3} and rates in mol dm^{-3} s^{-1}.

1 The equation for the reaction between ethanol and ethanoic acid to give ethyl ethanoate and water is shown below:

$$CH_3COOH + C_2H_5OH \rightleftharpoons CH_3COOC_2H_5 + H_2O$$

The concentrations of each of the species at equilibrium are controlled by the equilibrium law as shown below:

$$K_c = \frac{[CH_3COOC_2H_5][H_2O]}{[CH_3COOH][C_2H_5OH]}$$

Work out the units of the equilibrium constant, K_c.

2 The relationship between the concentrations of hydrogen ions, ethanoate ions and ethanoic acid molecules in an aqueous solution of ethanoic acid is given by the equation below:

$$K_a = \frac{[H^+][CH_3COO^-]}{[CH_3COOH]}$$ where K_a = acid dissociation constant of ethanoic acid.

Work out the units of the acid dissociation constant, K_a.

3 The rate of the reaction between iodine and propanone in acidic solution is given by the following equation:

$$\text{Rate} = k[CH_3COCH_3][H^+]$$

Work out the units of the rate constant, k.

4 The rate of reaction between aqueous sodium hydroxide and 2-bromo-2-methylpropane is given by the following equation:

$$\text{Rate} = k[\text{2-bromo-2-methylpropane}]$$

Work out the units of the rate constant, k.

5 When iodine in 1,1,1-trichloroethane (TCE) is shaken with water, two separate layers form; the two liquids are said to be **immiscible**. The concentration of iodine in each solvent is given by the following equation:

$$\frac{[I_2(\text{TCE})]}{[I_2(\text{aq})]} = K$$

Work out the units of the distribution coefficient, K.

Summary

You should now:

- know the rules for rearranging equations
- know the basic rule: always do the same thing to both sides of an equation
- be able to rearrange an equation
- be able to change the subject e.g. to be able to change $\frac{V_a M_a}{a} = \frac{V_b M_b}{b}$ into $M_a = \frac{V_b M_b a}{V_a b}$
- know that the SI units of length, mass and time are m, kg and s respectively
- be familiar with other units commonly used by chemists, e.g. mass measured in g, volume in cm^3
- appreciate the importance of using a consistent set of units e.g. in $n = cV$, if c is measured in mol dm^{-3} then V must be measured in dm^3
- be able to work out the units for a value calculated using an equation e.g. if in the equation $Q = m \times s \times \Delta T$, m is measured in g, s is measured in J g^{-1} K^{-1}, ΔT is measured in K, then Q is in J.

Answers to Try It Yourself Questions

Exercise 4.2.1, *p. 22*

1 $p = \frac{nRT}{V}$

2 $n = \frac{pV}{RT}$

3 $k = \frac{r}{[CH_3COCH_3][H^+]}$

4 $\Delta S = \frac{\Delta H - \Delta G}{T}$

5 $[H^+] = \frac{K_a[CH_3COOH]}{[CH_3COO^-]}$

Exercise 4.2.2, *p. 23*

1 $x = \frac{\sqrt{K_c}}{1 + \sqrt{K_c}}$

Exercise 4.3.2, *p. 25*

1 $R = \frac{pV}{nT} = \frac{1.00 \times 24.4}{1.00 \times 298}$

Rewriting this, including the units of each quantity gives the following:

$$R = \frac{1.00\ (\text{atm}) \times 24.4\ (\text{l})}{1.00\ (\text{mol}) \times 298\ (\text{K})}$$

Rewriting this with *only* the units gives the following:

$$R = \frac{\text{atm} \times \text{l}}{\text{mol} \times \text{K}} = \text{atm l mol}^{-1}\ \text{K}^{-1}$$

Exercise 4.3.3, *p. 27*

1 K_c has no units.
2 mol dm^{-3}
3 mol^{-1} dm^3 s^{-1}
4 s^{-1}
5 K has no units.

Appendix

The equilibrium law

If a mixture of ethanoic acid and ethanol is heated under reflux with some concentrated sulphuric acid as a catalyst, an equilibrium is established between these compounds and the products ethyl ethanoate and water. The equation for the reaction is:

$$CH_3COOH + C_2H_5OH \rightleftharpoons CH_3COOC_2H_5 + H_2O$$

The concentrations of each of the species at equilibrium are controlled by the equilibrium law as shown below:

$$K_c = \frac{[CH_3COOC_2H_5][H_2O]}{[CH_3COOH][C_2H_5OH]}$$ where K_c is the equilibrium constant.

If a mixture of ethanoic acid and ethanol, each of concentration 1 mol dm^{-3}, is brought to equilibrium, it can be shown that the concentration of ethyl ethanoate, x, will be given by the following equation:

$$K_c = \frac{x^2}{(1-x)^2}$$

We can easily show that this is so:

	CH_3COOH +	C_2H_5OH ⇌	$CH_3COOC_2H_5$ +	H_2O
Initial concentrations/ mol dm^{-3}	1	1	0	0
Equilibrium concentrations/ mol dm^{-3}	$1-x$	$1-x$	x	x

(If the concentration of ethyl ethanoate is x mol dm^{-3}, that of water will be the same because for every mole of ethyl ethanoate produced, 1 mole of water is formed. Each mole of ethyl ethanoate produced consumes 1 mole of ethanoic acid; producing x moles consumes x moles of ethanoic acid, leaving $1-x$. The same reasoning applied to ethanol produces the same equilibrium concentration.)

Substituting these values into the expression for K_c leads to:

$$K_c = \frac{x^2}{(1-x)^2}$$

Chapter 5

Ratios

After completing this chapter you should:

- *understand what is meant by a* **ratio**
- *understand the significance of two variables having a* **constant ratio**
- *understand what is meant by two variables being in* **direct proportion** *to one another*
- *be able to use ratios to make predictions.*

5.1 Introduction to ratios

Chemists often have to 'scale up' or 'scale down' the 'recipes' that they use. This may simply be because they need to make more (or less) than the recipe produces, or because a process is being transferred from the laboratory bench out into the factory, possibly via a pilot plant.

Sometimes all that is needed is to double all the quantities, or to scale up by a factor of 10. Often the numbers are not quite so simple. To carry out the calculations they need to use the idea of ratios.

Constant ratios

Look at these numbers.

Variable 1	1	2	3	4	5
Variable 2	2	4	6	8	10

Simply by looking, you can see that the numbers in row 2 are twice the values of those in row 1. A mathematical way of showing this is to calculate the value of:

$$\frac{\text{Variable 1}}{\text{Variable 2}}$$

for each of the columns. You will find that the value is constant at 0.5.

KEY FACT *The expression above is called a* **ratio**, *and the two sets of values are said to be* **in a constant ratio** *to each other.*

There are many examples of quantities that are in a constant ratio to each other in chemistry. Look at the results of a class experiment shown below.

A group of students each took a weighed sample of copper oxide, placed it in suitable apparatus and heated it in a stream of hydrogen gas. The copper oxide was reduced to copper. When the reaction was complete, and the copper had cooled down, it was weighed.

Some possible results are shown below.

Table 1 *Reduction of copper oxide*

Student	Dave	Louise	Ahmed	Kevin	Asma	Sharon
Mass of copper oxide taken/g	1.501	1.247	1.197	1.506	1.764	1.247
Mass of copper obtained/g	1.199	0.996	0.956	1.203	1.409	0.996
Mass of oxygen in copper oxide/g						
Ratio $\frac{\text{mass of copper}}{\text{Mass of oxygen}}$						

Try It Yourself

Exercise 5.1.1

1 Work out the values needed to complete the bottom row of Table 1 above. The mass of oxygen is the difference between the mass of copper oxide taken and the mass of copper remaining. When you have finished the exercise, you can check your answers with those on p. 35. You should have found a constant ratio of 3.97.

2 The results given above are idealised! In practice, it is unlikely that all students would obtain exactly the same ratio because of experimental error. For example, it is possible that not all of the oxygen will be removed in the experiment, so that at the end there will be a mixture of copper and unchanged copper oxide. How would this affect the ratio that would be obtained?

5.2 Making predictions

Once you know that two numbers (in this case the mass of copper and the mass of oxygen in a sample of copper oxide) are in a constant ratio to each other, you can use it to **predict** other sets of values. You can, for example, multiply *both* masses by any number you choose, *see Table 2*.

Table 2 *A constant ratio*

Mass of copper/g	mass of oxygen/g	Ratio $\frac{\text{mass of copper}}{\text{mass of oxygen}}$	
1.198	0.302	3.97	*Results of Dave's experiment*
2.396	0.604	3.97	*Both masses multiplied by 2*
3.594	0.906	3.97	*Both original masses multiplied by 3*

We have predicted that if we started with 3.594 g of copper, it would combine with 0.906 g of oxygen.

KEY FACT *You do not alter the value of a ratio if you multiply each value by the same number.*

Try It Yourself

Exercise 5.2.1

1 Work out the values needed to complete this table.

Mass of copper/g	mass of oxygen/g	Ratio $\frac{\text{mass of copper}}{\text{mass of oxygen}}$	
1.198	0.302	3.97	*Dave's experiment results*
			Both masses divided by 2
			Both original masses divided by 3

When you have completed it, you can check your answers with those on p. 35. You should have found that the ratio of the two masses remains constant.

Direct proportion

Let us now apply ratios to another chemical reaction.

The combustion of methane can be represented by the following chemical equation:

$$CH_4 + 2O_2 \rightarrow CO_2 + 2H_2O$$

This equation gives the following information about reacting quantities:

1 mole of methane reacts with 2 moles of oxygen to give 1 mole of carbon dioxide and 2 moles of water.

We can then convert these to masses using the relative molecular mass of each substance:

16 g of methane reacts with 64 g of oxygen to give 44 g of carbon dioxide and 36 g of water.

Methane will always react with oxygen in the ratio:

16 g of methane: 64 g of oxygen

Carbon dioxide and water will always be produced in the same ratio:

16 g of methane: 44 g of carbon dioxide: 36 g of water

If you start with a different mass of methane, the masses of oxygen, carbon dioxide and water will change, but they will always be in the ratio shown above (*see Table 3*).

Table 3 *Direct proportion*

Mass of methane/g	Mass of oxygen/g	Mass of carbon dioxide/g	Mass of water/g	
16.0	64.0	44.0	36.0	*Values from equation above*
32.0	128.0	88.0	72.0	*Multiplied by 2*
96.0	384.0	264.0	216.0	*Multiplied by 6*
144.0	576.0	396.0	324.0	*Multiplied by 9*
4.00	16.0	11.0	9.00	*Divided by 4*

Try It Yourself

Exercise 5.2.2

1 Work out the values needed to complete the following table.

HINT *In each case, you need to decide by what value the original mass of methane has been multiplied or divided. Then do the same to the masses of oxygen, carbon dioxide and water.*

Mass of methane/g	Mass of oxygen/g	Mass of carbon dioxide/g	Mass of water/g
16.0	64.0	44.0	36.0
8.0			
4.0			
24.0			

When you have completed the table, you can check your answers with those on p. 35.

KEY FACT *When two variables are in a constant ratio to each other, they are said to be in* ***direct proportion.***

In the combustion of methane, we can say:

- the mass of oxygen is *directly proportional* to the mass of methane
- the mass of carbon dioxide produced is *directly proportional* to the mass of methane
- the mass of water produced is *directly proportional* to the mass of methane

A number of other similar statements can be made. Can you think of at least one?

KEY FACT *When two numbers are directly proportional to one another, doubling one of them causes the other to double.*

Sometimes chemists have to complete ratio calculations that involve less obvious manipulations than simply doubling, halving or trebling the quantities.

Look again at the combustion of methane:

$$CH_4 \quad + \quad 2O_2 \quad \rightarrow \quad CO_2 \quad + \quad 2H_2O$$

This equation leads to the following ratios:

16 g of methane	reacts with	64 g of oxygen	to give	44 g of carbon dioxide	and	36 g of water.

It is fairly obvious that 32 g (2 × 16 g) of methane will require 128 g (2 × 64 g) of oxygen. The mass of oxygen needed to burn 50 g of methane is less obvious. However, the answer can be obtained in a series of steps.

Mass of methane/g	Mass of oxygen/g	
16	64	*From the equation*
$\frac{16}{16} = 1$	$\frac{64}{16} = 4$	*Divide by 16 to bring mass of methane down to 1.*
$1 \times 50 = 50$	$4 \times 50 = 200$	*Multiply by 50 (the required mass of methane).*

Have another look at this calculation to identify the steps. Remember that the question asks you to calculate the mass of oxygen required to burn 50 g of methane.

HINT

Start with the reacting masses from the equation.
Reduce the methane to 1 g by dividing by 16 (the mass of methane from the equation).
Divide the mass of oxygen by the same figure (16) to keep the ratio constant.
Multiply the mass of methane by 50 (the mass specified in the question).
Multiply the mass of oxygen by the same figure (50) to keep the ratio constant.

Try It Yourself

Exercise 5.2.3

1 Sulphur combines with oxygen to form sulphur dioxide as shown in the following equation.

$$S + O_2 \longrightarrow SO_2$$

(a) Calculate the mass of 1 mole of sulphur, 1 mole of oxygen and 1 mole of sulphur dioxide. A_r: O = 16, S = 32.
(b) Write down the masses of sulphur and oxygen required to produce 64 g of sulphur dioxide.
(c) Calculate the masses of sulphur and oxygen required to produce 100 g of sulphur dioxide.
(d) Calculate the mass of sulphur required to produce 100 tonnes of sulphur dioxide (1 tonne = 1000 kg).

Summary

You should now:

- know that if $\frac{a}{b}$ is a constant for all values of a and b they are said to be in a constant ratio
- know that if $\frac{a}{b}$ is a constant for all values of a and b they are said to be directly proportional to each other
- know that multiplying a and b by the same factor does not alter the ratio
- be able to predict the value of b for a given value of a.

Answers to Try It Yourself Questions

Exercise 5.1.1, *p. 31*

1

Student	Dave	Louise	Ahmed	Kevin	Asma	Sharon
Mass of copper oxide taken/g	1.501	1.247	1.197	1.506	1.764	1.247
Mass of copper obtained/g	1.199	0.996	0.956	1.203	1.409	0.996
Mass of oxygen in copper oxide/g	0.302	0.251	0.241	0.303	0.355	0.251
Ratio $\frac{\text{mass of copper}}{\text{mass of oxygen}}$	3.97	3.97	3.97	3.97	3.97	3.97

2 If the copper oxide had not all been reduced to copper, the mass of oxygen would be smaller and the mass of copper would be bigger. The ratio

$$\frac{\text{mass of copper}}{\text{mass of oxygen}}$$

would therefore have been greater than 3.97.

Exercise 5.2.1, *p. 32*

1

Mass of copper/g	Mass of oxygen/g	Ratio $\frac{\text{mass of copper}}{\text{mass of oxygen}}$	
1.198	0.302	3.97	*Dave's experiment results*
0.599	0.151	3.97	*Both masses divided by 2*
0.240	0.0604	3.97	*Both original masses divided by 5.*

Exercise 5.2.2, *p. 33*

1

Mass of methane/g	Mass of oxygen/g	Mass of carbon dioxide/g	Mass of water/g	Scale factor*
16.0	64.0	44.0	36.0	
8.0	32.0	22.0	18.0	0.5
4.0	16.0	11.0	9.0	0.25
24.0	96.0	66.0	54.0	1.5

*Scale factor = $\frac{\text{mass of methane in this now}}{\text{mass of methane originally (16 g)}}$

All the values in this row are multiplied by this scale factor.

Exercise 5.2.3, *p. 34*

1 (a) 32 g of sulphur, 32 g of oxygen, 64 g of sulphur dioxide
(b) 32 g of sulphur and 32 g of oxygen
(c) 50 g of sulphur and 50 g of oxygen
(d) 50 tonnes of sulphur

Exam Questions

Exam type questions to test understanding of Chapters 1–5

1 (a) 1.00 g of sodium hydroxide are dissolved to give 250.0 cm^3 of solution. Calculate the concentration of the solution in mol dm^{-3}.

(b) 25.0 cm^3 of this solution of sodium hydroxide was titrated against hydrochloric acid using methyl orange as indicator. The end point occurred after addition of 22.8 cm^3 of the hydrochloric acid. Calculate the concentration of the hydrochloric acid in mol dm^{-3}

(A_r: H = 1.01, O = 16.0, Na = 23.0.)

HINT

In an examination you would be expected to know the relationship between the volume and concentrations of the reagents as shown below:

$\frac{M_aV_a}{a} = \frac{M_bV_b}{b}$ where M_a and V_a are the concentration and volume of reagent A and a moles of A react with b moles of B.

You would also be expected to know that hydrochloric acid reacts with sodium hydroxide in a ratio of 1 mole : 1mole.

2 A sample of gas was found to occupy 30.1 cm^3 at 290 K and 1.02×10^5 Pa.

(a) Calculate the number of moles of gas in the sample.
(Gas Constant, $R = 8.31$ J mol^{-1} K^{-1}.)

(b) Calculate the volume the gas would occupy at 373 K and 1.02×10^5 Pa.

HINT

In an examination you would be expected to know the relationship:

$pV = nRT$ Where P is pressure, V is volume and Y the temperature of n moles of an ideal gas.

You then need to make n *the subject of the formula.*

3 0.105 g of a gas was found to occupy a volume of 35.9 cm^3 at 295 K and 1.01×10^5 Pa. Calculate the relative molecular mass of the gas. (Gas Constant, R, = 8.31 J mol^{-1} K^{-1}.)

4 The lattice enthalpy of sodium chloride = −771 kJ mol^{-1}

The hydration enthalpy of the sodium ion = −406 kJ mol^{-1}

The hydration enthalpy of the chloride ion = −364 kJ mol^{-1}

Use these data to calculate the enthalpy of solution of sodium chloride.

HINT

In an examination you would be expected to be able to draw an enthalpy cycle and work out the relationship:

$\Delta H_{solution} = -\Delta H_{lattice} + \Delta H_{hydration} Na^+ + \Delta H_{hydration} Cl^-$

5. The lattice enthalpy of potassium chloride = −701 kJ mol^{-1}

The enthalpy of solution of potassium chloride = +17.2 kJ mol^{-1}

The hydration enthalpy of the chloride ion = −364 kJ mol^{-1}

Use these data to calculate the hydration enthalpy of the potassium ion.

6 The enthalpy of formation of ethane = −847 kJ mol^{-1}

The enthalpy of formation of carbon dioxide = −394 kJ mol^{-1}

The enthalpy of formation of water = −286 kJ mol^{-1}

Use these data to calculate the enthalpy of combustion of ethane.

HINT

In an examination you would be expected to draw an energy cycle and derive the relationship:

$\Delta H_c(\text{ethane}) = 2\Delta H_f(CO_2) + 3\Delta H_f(H_2O) - \Delta H_f(C_2H_6)$

7 The enthalpy of combustion of propane = -2220 kJ mol^{-1}

The enthalpy of formation of carbon dioxide = -394 kJ mol^{-1}

The enthalpy of formation of water = -286 kJ mol^{-1}

Use these data to calculate the enthalpy of formation of propane.

8 The initial rates of decomposition of sulphur dichloride dioxide (SO_2Cl_2) into sulphur dioxide and chlorine for different initial concentrations are shown in the table below.

Experiment	Initial concentration of SO_2Cl_2/mol dm^{-3}	Initial rate of decomposition/ mol dm^{-3} s^{-1}
1	0.500	1.50×10^{-4}
2	0.250	7.50×10^{-5}
3	0.125	3.75×10^{-5}

(a) Determine the order of this reaction.

(b) Calculate the value of the rate constant and state the units.

9 The rate of reaction between A and B at constant temperature was studied, and the following values for initial rates were obtained.

Experiment	Initial [A]/ mol dm^{-3}	Initial [B]/ mol dm^{-3}	Initial rate/ mol dm^{-3} s^{-1}
1	0.010	0.010	1.0×10^{-6}
2	0.010	0.020	4.0×10^{-6}
3	0.020	0.030	9.0×10^{-6}

(a) Use the data to decide which of the following statements is/are correct:

(i) Rate of reaction does not depend on the concentration of A.

(ii) Rate is directly proportional to [A].

(iii) Rate is directly proportional to $[A]^2$.

(iv) Rate of reaction does not depend on [B].

(v) Rate is directly proportional to [B].

(vi) Rate is directly proportional to $[B]^2$.

(b) Which of the following is the correct rate equation for this reaction?

(i) Rate = $k[A][B]$ (ii) Rate = $k[A][B]^2$

(iii) Rate = $k[A]^2[B]$ (iv) Rate = $k[A]^2[B]^2$

(v) Rate = $k[A]^2$ (vi) Rate = $k[B]^2$

(c) Use your results to calculate the value of the rate constant, k, including units. From this calculate the initial rate for initial concentrations of A and B of 0.15 and 0.25 mol dm^{-3} respectively.

10 Iodine chloride, ICl, reacts with hydrogen to produce hydrogen chloride and iodine. The initial rates of reaction for different initial concentrations of reactants are shown in the table below.

Experiment	Initial [ICl]/ mol dm^{-3}	Initial [H_2]/ mol dm^{-3}	Initial rate of reaction/ mol dm^{-3} s^{-1}
1	0.10	0.15	0.0473
2	0.20	0.15	0.0945
3	0.30	0.30	0.567

Use the above results to determine the order of reaction with respect to iodine chloride and hydrogen, and hence the rate equation. Calculate the rate constant.

11 Iodine reacts with propanone in acidic solution to give iodopropanone. The initial rate of this reaction was measured for different initial concentrations of the reagents. The results are shown in the table below.

Experiment	Initial [CH_3COCH_3]/ mol dm^{-3}	Initial [I_2]/ mol dm^{-3}	Initial [H^+]/ mol dm^{-3}	Initial rate of reaction/ mol dm^{-3} s^{-1}
1	0.11	0.60	0.59	0.000 11
2	0.22	0.60	0.59	0.000 22
3	0.11	0.60	1.79	0.000 33
4	0.22	0.30	1.20	0.000 44

Use these results to deduce the rate equation and to obtain a value for the rate constant.

12 An organic compound was found on analysis to have the following composition by mass: C = 92.3%, H = 7.70%. Work out the empirical formula of the compound (A_r: H = 1.01, C = 12.0).

13 An oxide of iron was found to contain 70.0% iron by mass. Work out the empirical formula of the oxide (A_r: O = 16.0, Fe = 55.8).

14 In a saturated solution of calcium sulphate at 298 K, the relationship between the concentrations of the two ions is given by the following expression:

$$[Ca^{2+}(aq)][SO_4^{2-}(aq)] = 2.4 \times 10^{-5}\ mol^2\ dm^{-6}$$

(a) Calculate the concentration of calcium ions in a saturated solution at 298 K.

(b) Use your answer to part (a) to calculate the solubility of calcium sulphate in water in g dm^{-3} at this temperature.

(c) Solid calcium sulphate is shaken with 0.100 mol dm^{-3} sulphuric acid. Calculate the maximum concentration of calcium ions that can be achieved in the solution. You can assume that the concentration of sulphate ions caused by the calcium sulphate can be neglected (A_r: Ca = 40.1, S = 32.1, O = 16.0).

15 In a saturated solution of silver sulphate at 298 K, the relationship between the concentrations of the two ions is given by the following expression:

$$[Ag^+(aq)]^2[SO_4^{2-}(aq)] = 1.7 \times 10^{-5}\ mol^3\ dm^{-9}$$

Calculate the solubility of silver sulphate in g dm^{-3} in water at 298 K (A_r: Ag = 108, S = 32.1, O = 16.0).

16 A mixture of hydrogen and iodine reacts to form an equilibrium with hydrogen iodide as shown in the following reaction:

$$H_2(g) + I_2(g) \rightleftharpoons 2HI(g)$$

The equilibrium constant, K_c is given by the following expression:

$$K_c = \frac{[HI(g)]^2}{[H_2(g)][I_2(g)]}$$

The value of K_c at a particular temperature was found to be 46.0. Calculate the concentration of hydrogen iodide in an equilibrium mixture at this temperature, if the equilibrium concentrations of hydrogen and iodine are 2.64 and 2.12 mol dm^{-3} respectively.

Chapter 6

Graphs

After completing this chapter you should:

- *know that graphs can be used to display or interpret data*
- *know that graphs can take different forms scatter graphs, e.g. line graphs, bar charts and pie charts*
- *understand the significance of the gradient and intercept in straight-line graphs.*

6.1 Introduction to graphs

Graphs can be used either to *display* data in a convenient way, or to *interpret* data that is to derive further information from it. There are many different types of graph, including scatter graphs (*Fig. 1*), bar charts (*Fig. 2*), pie charts (*Fig. 3*) line graphs (*Fig. 4*) and curves (p. 46). In this chapter, we will be looking at line graphs and curves; we shall look at other types in Chapter 7.

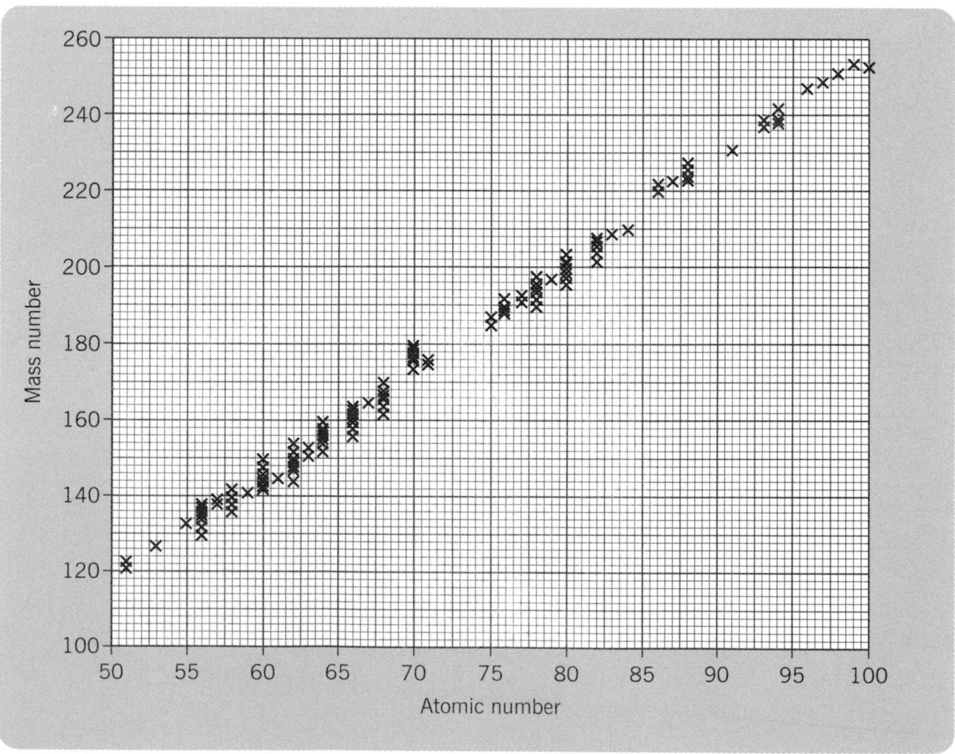

Fig. 1 *Scatter graph of stable nuclei.*

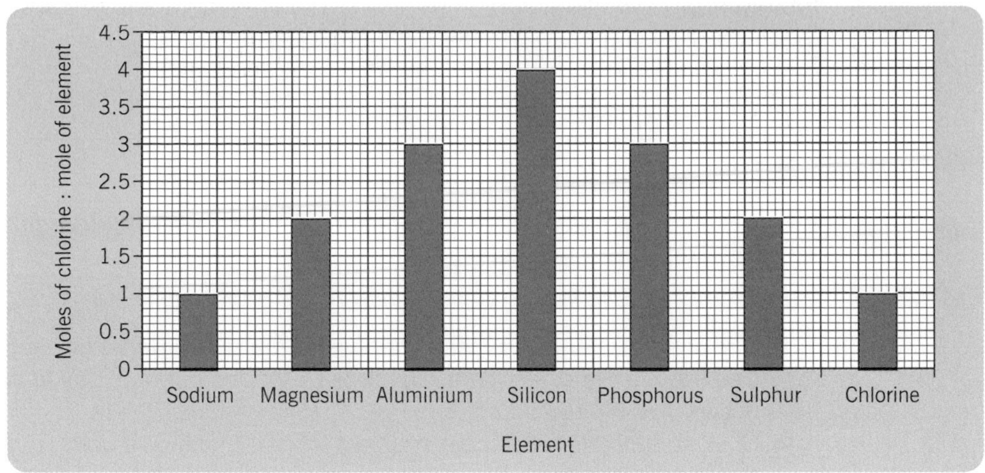

Fig. 2 *Formulae of chlorides of Period 3.*

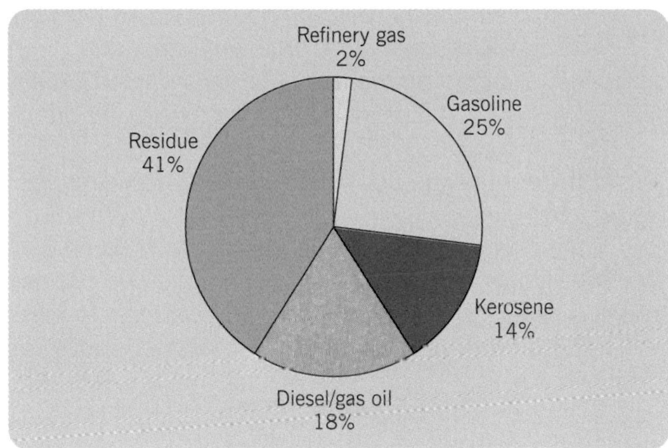

Fig. 3 *Pie chart showing the composition of a sample of crude oil.*

6.2 Linear (straight-line) graphs

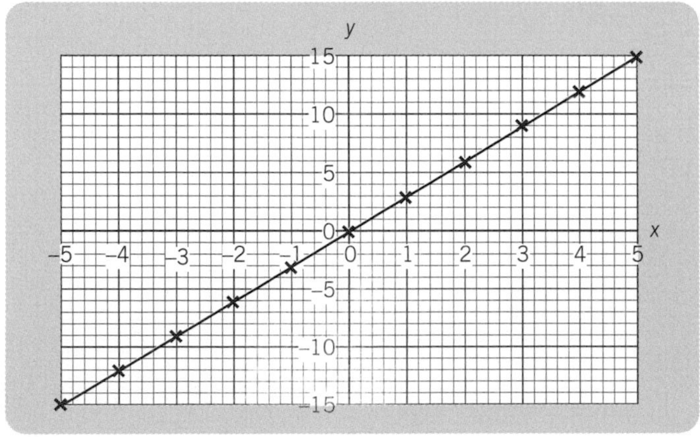

Fig. 4 *Graph of* $y = 3y$.

Fig. 4 shows a graph of $y = 3x$ plotted for values of x between −5 and +5.

Although this is a simple graph to plot, if you are not sure how it is done, you might find it useful to copy and complete the following table.

$x =$	−5	−4	−3	−2	−1	0	1	2	3	4	5
$3x =$											

You can then use this to plot the graph.

REMEMBER *Remember the following points when plotting graphs:*

- *Choose a scale which makes full use of the graph paper. A useful rule is that if you can double the scale (either in the x- or y-direction) and still fit all the points on to the paper, you should do so.*
- *Choose a convenient scale such as 1 cm = 1 unit, or 2 units, or 5 units, or 10 units. Avoid difficult scales e.g. 1 cm = 3 units; it is easy to make a mistake with such a scale.*
- *Label each axis. In the case of this graph, labelling them x and y is appropriate. For most graphs plotted in chemistry, you will be plotting something that has been measured, such as mass, temperature, time, volume, etc.*
- *You should normally indicate on the axes the units of the quantity being plotted; the only exception to this is for the small number of measured quantities that have no units.*
- *Plot the points as accurately as possible – certainly to within the accuracy to which they are measured or given.*
- *In most cases, it is appropriate to join up the points in some way. For most graphs in chemistry this will involve drawing a straight line, the line of best fit (see p. 75) or a smooth curve. It is rarely appropriate in chemistry simply to join up the points by going from one to the other (usually referred to as 'dot to dot', after children's drawing books!).*

Notice that the graph is a *straight line*, passing through *the origin* (the point where $x = 0$ and $y = 0$).

Figs. 5 and 6 show graphs of $y = 4x$ and $y = 5x$.

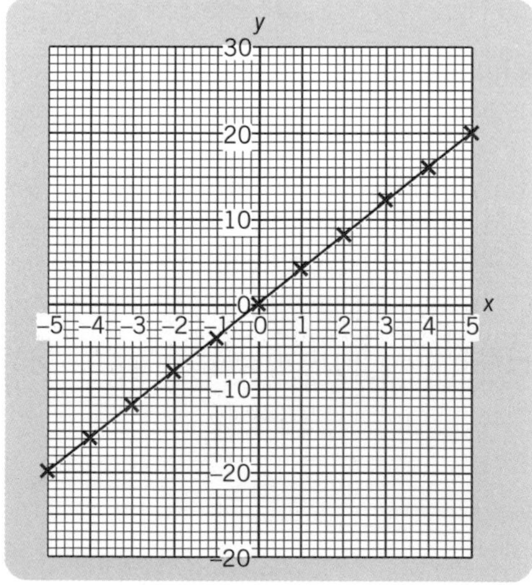

Fig. 5 *Graph of y = 4x.*

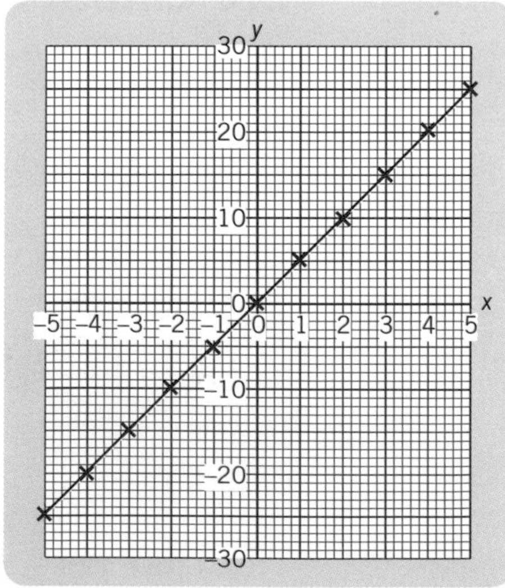

Fig. 6 *Graph of y = 5x.*

In what way are the three graphs (*Figs. 4, 5 and 6*) similar? In what way do they differ?

Gradients and intercepts

You should notice that the only difference between the three graphs is the *angle* at which the line rises. This is known as the **gradient**. It can be measured by drawing a triangle and measuring the increase in both x and y, usually given the symbols Δx and Δy respectively.

The gradient is then the ratio of these quantities:

$$\text{Gradient} = \frac{\text{increase in } y}{\text{increase in } x} = \frac{\Delta y}{\Delta x}$$

Fig. 7 shows how this can be done for the graph of $y = 3x$, although of course you may have drawn a different triangle.

$$\frac{\Delta y}{\Delta x} = \frac{12 - (-10)}{4 - (-3.3)} = \frac{22}{7.3} = 3.0$$

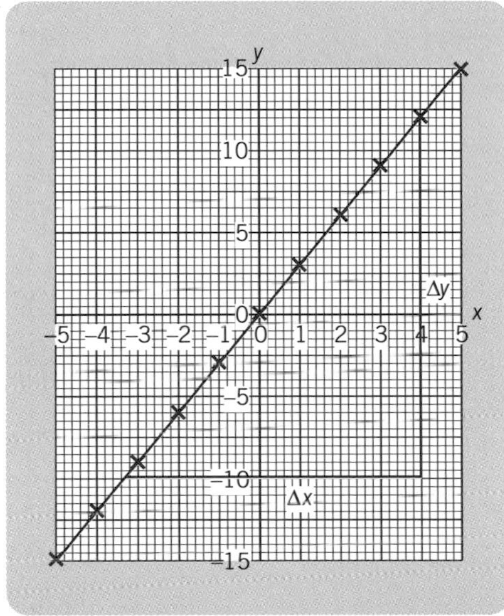

Fig. 7 *Graph of* y = 3x *showing the gradient.*

KEY FACT *When measuring gradients in this way, you should remember the following points:*

- *The sides of the triangle must be horizontal and vertical (parallel to the x and y axes respectively).*
- *Always draw as big a triangle as you conveniently can. This enables you to measure the length of each side accurately. If you use a small triangle, the measurements could be inaccurate, which means that the value for the gradient will be inaccurate (as it will be calculated using inaccurate values).*

Question: Measure the gradients of the graphs of $y = 4x$ and $y = 5x$ (*Figs. 5 and 6*).

Answer: You should find that they come to 4 and 5 respectively.

The fact that the graph of $y = 3x$ has a gradient of 3 is not a coincidence! The graph of $y = 4x$

has a gradient of 4; similarly, the graph of $y = 5x$ has a gradient of 5.

Note that these gradients are described as *positive* (y increases as x increases). Gradients sloping in the opposite direction (y decreases as x increases) are described as *negative* or have negative values (*see Fig. 8*).

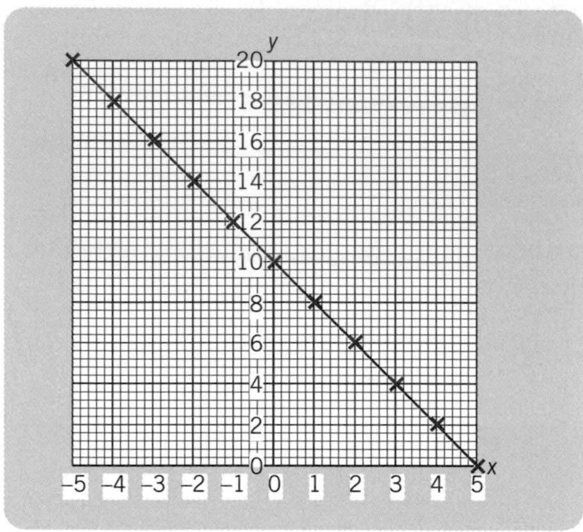

Fig. 8 *Graph with a negative gradient* ($y = 2x + 10$)*.*

This leads to the following general results.

> **KEY FACT** *A graph of* $y = mx$ *(where* m *is simply a number) will:*
>
> - *be a straight line*
> - *have a gradient m*
> - *pass through the origin.*

Figs. 9 and 10 show graphs of $y = 6x + 1$ and $y = 2x - 3$ respectively.

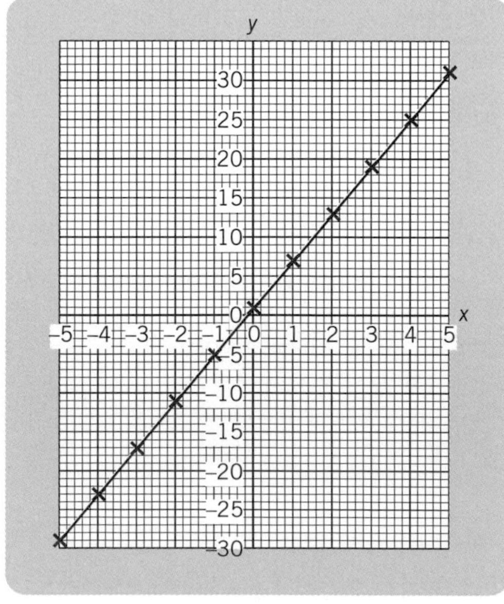

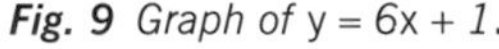
Fig. 9 *Graph of* $y = 6x + 1$*.*

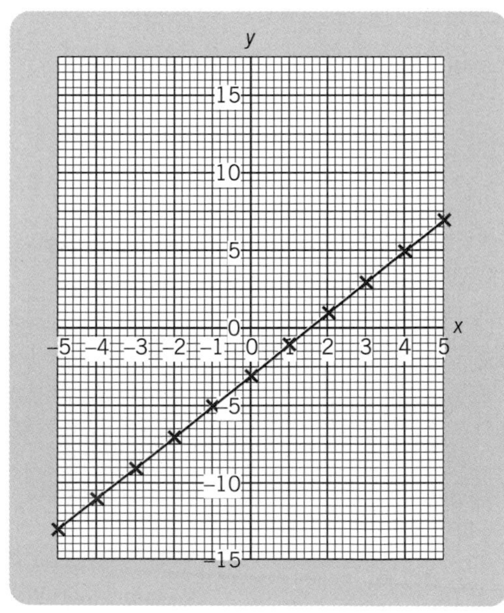

Fig. 10 *Graph of* $y = 2x - 3$*.*

Question: Predict the gradient of each of these, and then check your answer by measuring it.

Answer: You should predict and find gradients of 6 and 2 respectively.

Question: Now find the point where each graph cuts the y-axis (the value of y when $x = 0$). This is known as the ***y*-intercept**.

Answer: You should have found values of 1 and –3 respectively.

Now have another look at the original equations. Can you see a pattern?

Try It Yourself

Exercise 6.2.1

1 Try applying your pattern to complete the following table:

	Graph	Predicted gradient	Predicted y-intercept
(a)	$y = 6x + 4$		
(b)	$y = x - 2$		
(c)	$y = 7 - 4x$		

The graphs are shown below (*Figs. 11–13* respectively), so you can check your predictions. (See p. 47 for answers.)

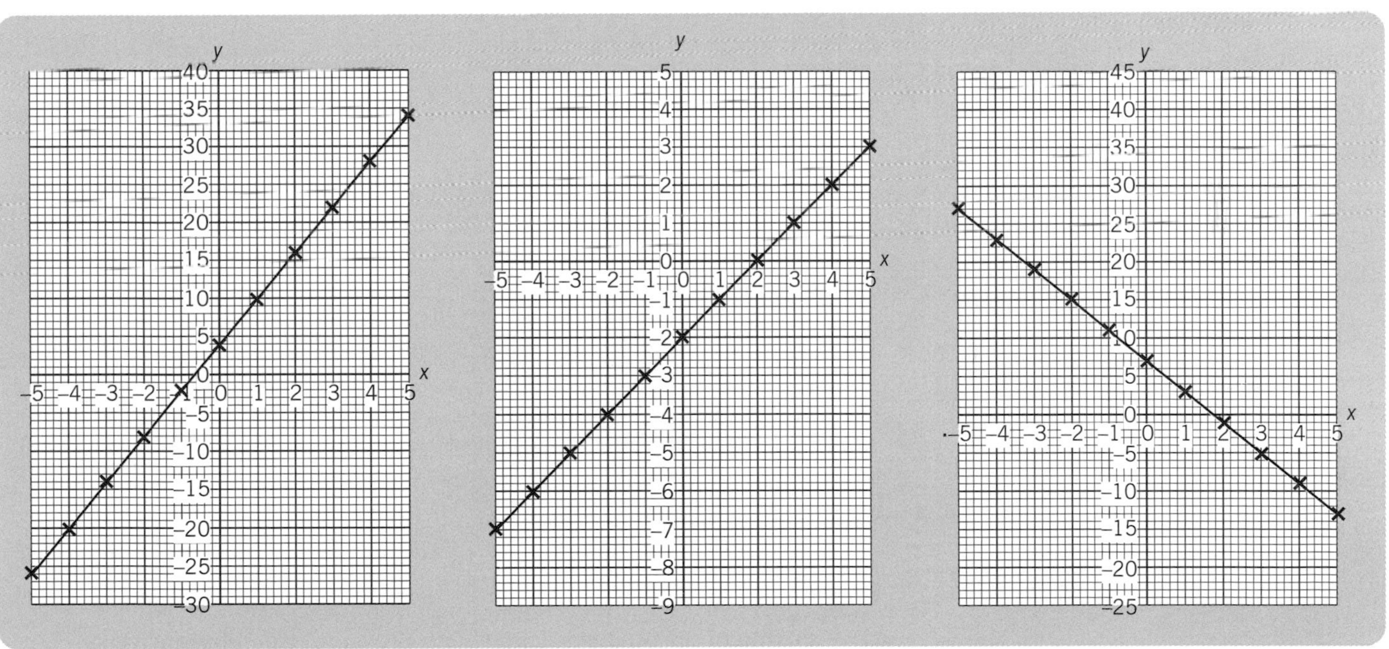

Fig. 11 *Graph of* $y = 6x + 4$.

Fig. 12 *Graph of* $y = x - 2$.

Fig. 13 *Graph of* $y = 7 - 4x$.

KEY FACT *A graph of* $y = mx + c$ *(where* m *and* c *are numbers) will have a* y- *intercept of* c.

6.3 Non-linear graphs (curves)

The most important graphs in chemistry will usually be straight lines However, a small number will lead to curves. *Fig. 14* shows how the area of a circle varies with radius the relationship is: area = $\pi \times (\text{radius})^2$.

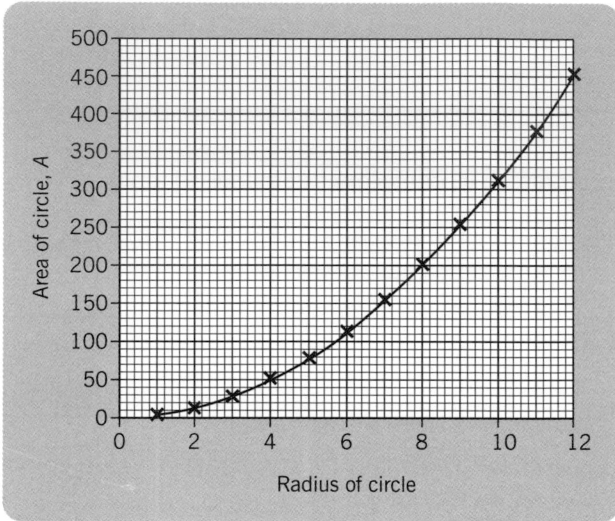

Fig. 14 *Graph of* $A = \pi r^2$.

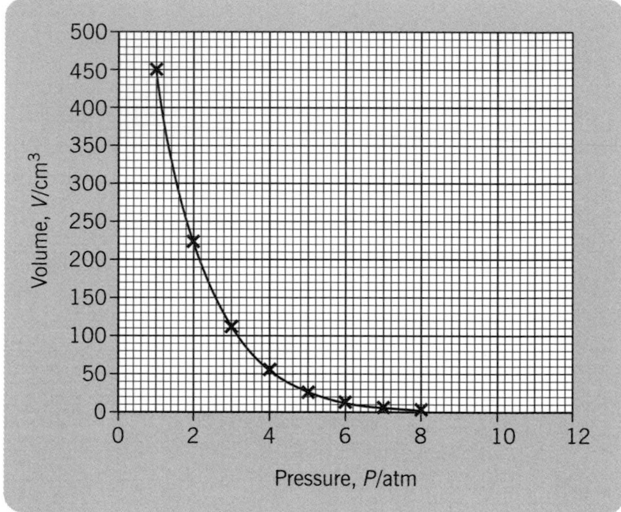

Fig. 15 *Graph of* $V = \dfrac{k}{\text{Pressure}}$.

Fig. 15 shows how the volume of (ideal) gases varies with pressure; the relationship is: volume = k/pressure, where k is a constant which depends on the amount of gas present and the temperature.

It is not immediately obvious from *Fig. 15* that the curve never actually reaches the x-axis. However, the curve comes closer and closer to the axis but never reaches it, even at very high values of x. This feature is known as an **asymptote**.

Summary

You should now know:

- graphs can be used to display or interpret data
- graphs can take different forms, e.g. scatter grams, line graphs, bar charts and pie charts
- a straight-line graph can be represented by $y = mx + c$ where m is the **gradient** and c is the y-**intercept** (the value where the graph cuts the y-axis)

Answers to Try It Yourself Questions

Exercise 6.2.1, *p. 45*

	Graph	Predicted gradient	Predicted y-intercept
(a)	$y = 6x + 6$	6	4
(b)	$y = x - 2$	1	−2
(c)	$y = 7 - 4x$	−4	7

Chapter 7

Using graphs

After completing this chapter you should:

- *know that graphs can be used to display data*
- *know the different types of graph that can be used*
- *know which type is appropriate to any given set of data*
- *know how graphs can be used to obtain data*
- *know how to use graphs to find relationships*
- *recognise spectra as a type of graph*
- *be able to read values from spectra*
- *be able to interpret frequency distribution curves.*

7.1 Displaying data

Graphs can be used in a number of ways; they can be used to display data in a form that helps to show patterns. For example, a plot of first ionisation energies against atomic number shows that they are *periodic*; that there is a repeating pattern. The pattern is also a piece of evidence for the existence of *electron shells* (or energy levels).

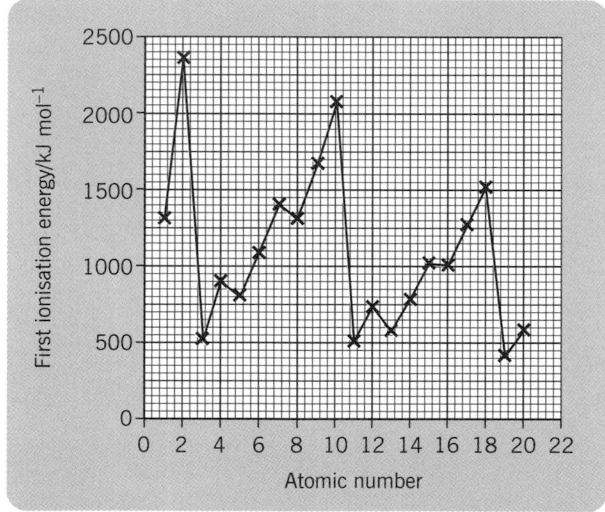

Fig. 1 *Graph of first ionisation energy vs. atomic number.*

Dot-to-dot graphs like *Fig. 1* are useful for showing-up patterns. Strictly speaking this is not a *graph* because the quantities are not *continuous*, e.g. 12.5 is an impossible value for an atomic number. Some would prefer to use a bar chart for such a situation. This can have advantages, e.g. the bars can be different colours to show some aspect of the pattern.

Fig. 2 shows the same data in bar chart form, with the values for the noble gases coloured. However it is common for chemists to use the dot-to-dot graph. *Fig. 2* is an example of a bar chart, as all the bars cover the same range (each represent one unit of atomic number). If the bars do not cover equal ranges, the graph is called a **histogram**.

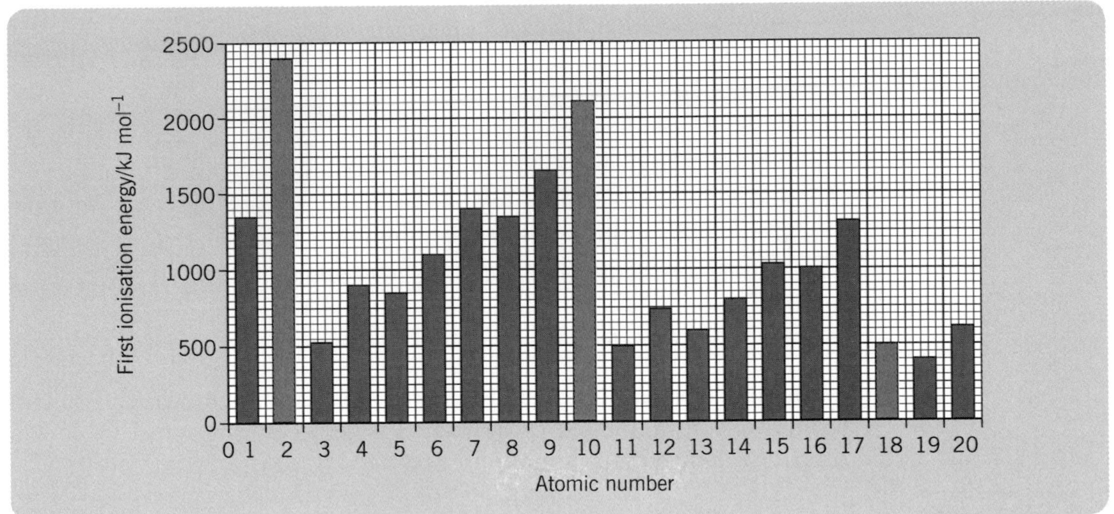

Fig. 2 *Bar chart of first ionisation energy vs. atomic number.*

7.2 Comparing quantities – pie charts

The usual way of displaying data so that one quantity can be compared with another is thc pie chart. For example, the total energy required for the conversion of aluminium atoms into Al^{3+} ions is 5138 kJ mol^{-1}. This is made up of three components as shown in the table.

Change	Energy/kJ mol^{-1}
$Al(g) \longrightarrow Al^{+}(g) + e^{-}$	577
$Al^{+}(g) \longrightarrow Al^{2+}(g) + e^{-}$	1816
$Al^{2+}(g) \longrightarrow Al^{3+}(g) + e^{-}$	2745

From the numbers we can see how the values progress, but presenting them as a pie chart (*Fig. 3*) shows not only this but also their relative importance.

In a pie chart, the size of each 'slice' or '**sector**' (to use the correct term) is proportional to the data which it represents. In this case, the first ionisation energy (577 kJ mol^{-1}) makes up 11.2%, and so it should occupy 11.2% of the circle. This is done in terms of angles. The sum of the angles of all the sectors must add up to 360° so the first ionisation energy should be represented by an angle of:

$$\frac{11.2}{100} \times 360° = 40°$$

Pie and bar charts are used only to *display* data rather than to interpret it – there are no slopes or intercepts to measure in them. Pie charts can be conveniently computer generated.

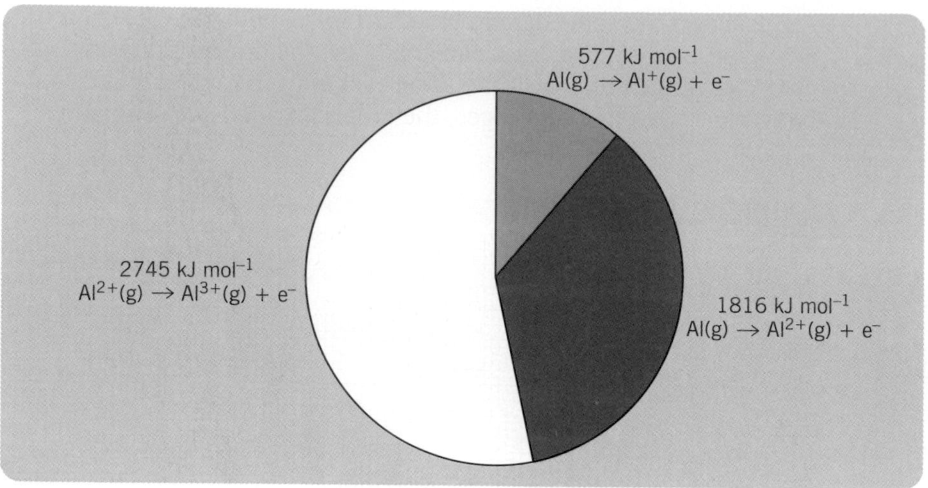

***Fig. 3** Pie chart of first three ionisation energies of aluminium.*

You may have access to statistics or graphics software which you can use for this purpose; alternatively, a number of standard spreadsheet programs enable you to do this. They do, however, have their limitations, and need to be used with care!

When using computer software, choose very carefully which option you select for displaying the data. For example, it is very easy to end up with a 'dot-to-dot' graph where a smooth curve would be more appropriate. Take care also to choose a display option which spaces the values on the *x*-axis correctly. This is often called a 'scatter' plot or a 'scattergram'. Other options space the *x*-values equally irrespective of their values, so that 0, 1, 2, 3, 4 is plotted in the same way as 0, 1, 2, 3, 10.

Try It yourself

Exercise 7.2.1

For each of the following, suggest the most suitable type of graph for displaying the data, giving your reasons. There may be alternative answers to some questions.

1 The pressure of a fixed volume of a gas at different temperatures.

2 How the density of liquid hydrocarbons at s.t.p. varies with the number of carbon atoms per molecule.

3 The quantity of different polymers manufactured in the UK during the past 12 months.

7.3 Obtaining data from graphs

Fig. 4 is a graph of initial rate of decomposition of hydrogen peroxide against its initial concentration.

The fact that it is a straight line tells you that the rate and the concentration are related by the following equation:

Rate = $k[H_2O_2]$ (See p. 44)

The gradient of the graph gives you the value of k.

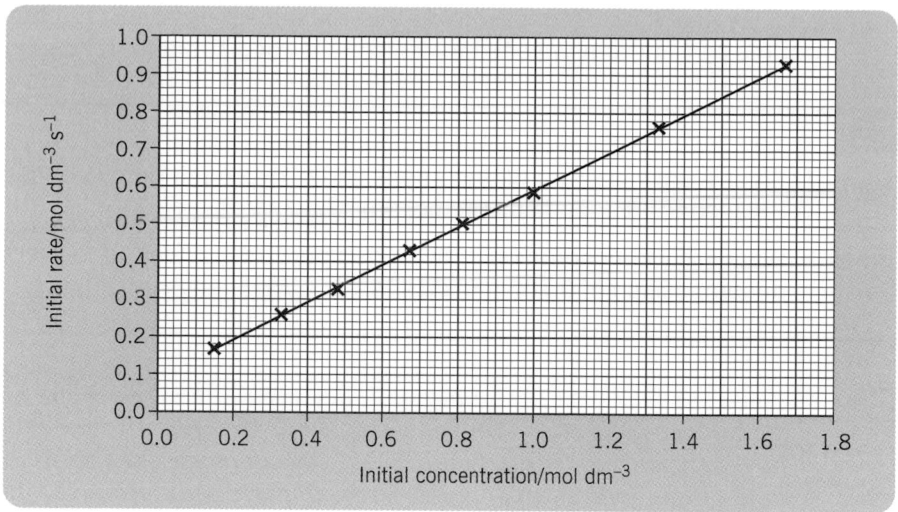

Fig. 4 *Decomposition of H_2O_2.*

Interpolation

You can use this graph to find out what the initial rate of reaction would be for any initial concentration of hydrogen peroxide (at the temperature at which the original readings were taken). You simply find the value you want on the x-axis, draw a *vertical* line up to the curve and then a *horizontal* line to the y-axis (see *Fig. 5*).

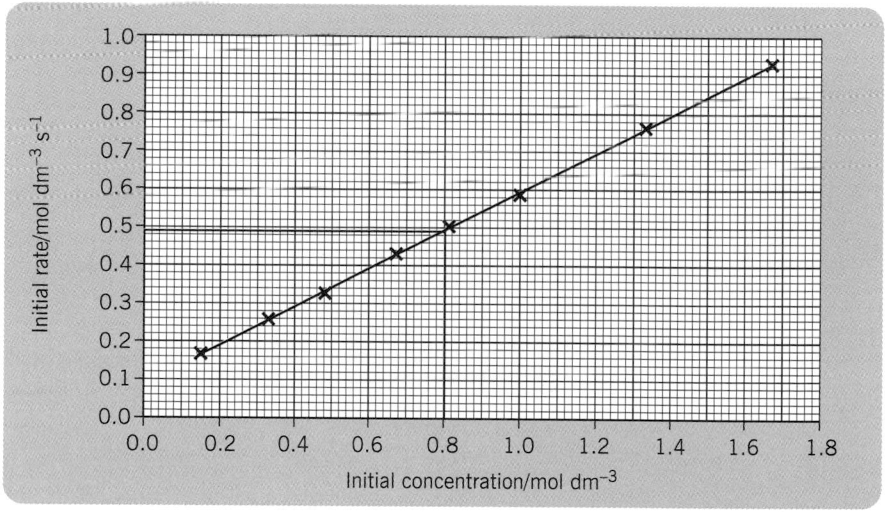

Fig. 5 *Finding the initial rate of reaction.*

This technique, where you use the graph to find information under conditions *between* those at which you have taken measurements, is known as **interpolation**.

You should be able to see for yourself how you could use the graph to predict the initial concentration required to achieve a particular initial rate of decomposition. Whilst it is unlikely that you would need to do this, an industrial chemist might use the technique to find out the concentration needed to achieve a particular rate of reaction in a chemical plant.

Try It Yourself

Exercise 7.3.1

1 What is the initial rate of decomposition of hydrogen peroxide (see *Fig. 4*) when the initial concentration is 0.900 mol dm^{-3}?

2 What initial concentration of hydrogen peroxide is needed to produce an initial rate of decomposition of 0.700 mol dm^{-3} s^{-1}?

WARNING!

You must take care to use interpolation only when the results have some significance. Fig. 1 *(p. 48) shows the variation of ionisation with atomic number however, the data is* not *continuous. You could use the technique of interpolation to read off the ionisation energies between two points. However, this would be meaningless, as has already been said, there is no element with, for example, an atomic number of 12.5.*

Extrapolation

A similar technique can be used to find values *outside* the range for which measurements are made. In this case, the straight line (or in other cases, smooth curve) is simply extended. *Fig. 6* shows data for enthalpy of combustion of a number of alkanes up to and including decane, $C_{10}H_{22}$. To estimate the enthalpy of combustion of dodecane, $C_{12}H_{26}$, you would simply extend the graph, and then read off the value in the way already described.

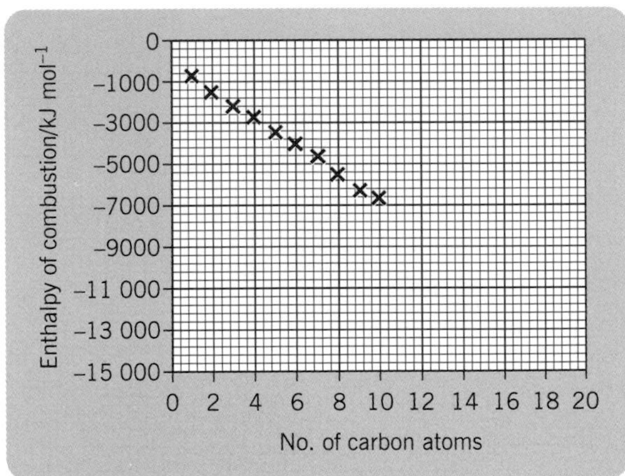

Fig. 6 Graph of enthalpy of combustion of alkanes against number of carbon atoms.

When the graph is extended outside the range of points in this way, it is known as **extrapolation**.

Try It Yourself

Exercise 7.3.2

1 Use the graph in the way described above to estimate the enthalpy of combustion of dodecane, $C_{12}H_{26}$.

2 Use the graph to estimate the enthalpy of combustion of eicosane, $C_{20}H_{42}$.

WARNING!

Extrapolation must be used with care. You need to be confident that the trend displayed on your graph continues beyond the end of the graph. For example, by extrapolating you make the assumption that the same conditions apply outside the measured range. This is a reasonable assumption in question 2 above, so that the estimate for the enthalpy of combustion of eicosane is likely to be reasonable.

However, this is not always the case.

Fig. 7 shows how the volume of a fixed mass of steam at constant pressure varies over the temperature range 400–600 K.

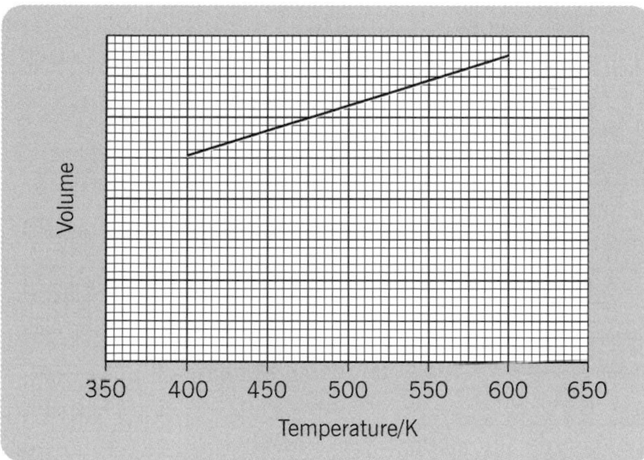

Fig. 7 *A fixed mass of steam at constant pressure.*

It could be extrapolated to estimate the volume at 300 K. However, as the steam would have condensed to give liquid water by then, extrapolation would produce a meaningless result. *Fig. 8* shows what really happens.

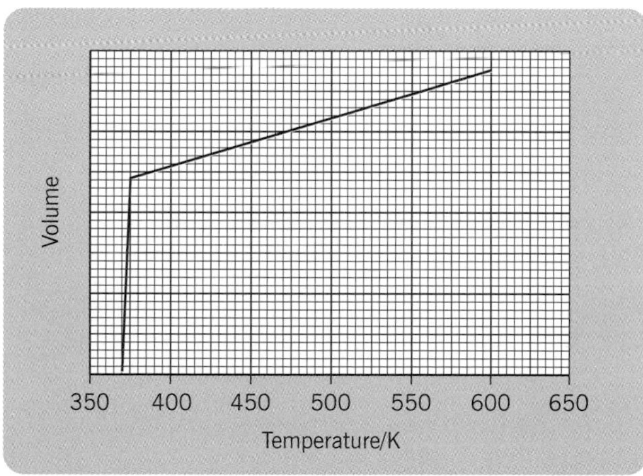

Fig. 8 *A fixed mass of H_2O at constant pressure.*

If you look back to *Fig. 1* (p. 48) it is unlikely that you can obtain a sensible estimate for the first ionisation energy of element 21 simply by extending the graph. In fact, from your knowledge of periodicity, you may realise that the most likely trend after element 20 is a *decrease* in ionisation energy, as happens after elements 4 and 12.

7.4 Establishing relationships

Graphs can also be used to establish the relationship between two properties. For example, whilst the rate of a chemical reaction increases with increasing concentration of reactants, in some cases doubling the concentration doubles the rate of reaction, whilst in other cases the rate increases by a factor of four. We can use graphs to find the *relationship* between rate and concentration of reactants.

Graphs can also be used to work out a *value* from experimental data. For example, the **activation energy** of a reaction can be calculated from measurements of rate of reaction at different temperatures.

Using straight-line graphs to find how two properties are related

The table below shows some (idealised!) results from an experiment in which the initial rate of a chemical reaction was measured at different initial concentrations.

These data are plotted on *Fig. 9*.

Initial [A]/ mol dm^{-3}	Initial rate of reaction/ mol dm^{-3} s^{-1}
0.100	0.500
0.200	1.00
0.300	1.50
0.400	2.00
0.500	2.50
0.600	3.00
0.700	3.50
0.800	4.00
0.900	4.50

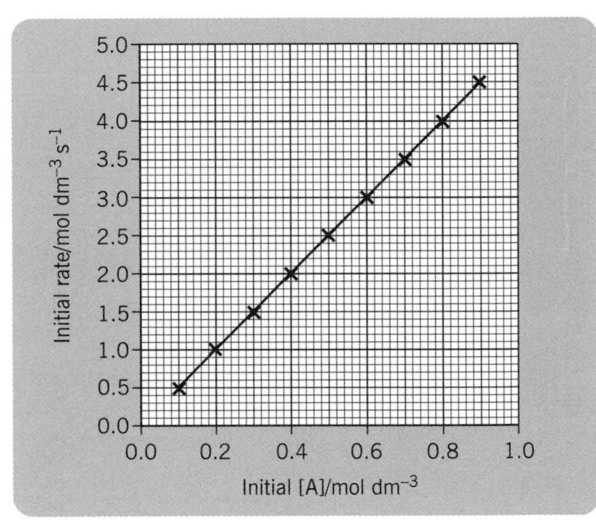

Fig. 9 *Rate of reaction against concentration of A.*

The fact that the graph is a straight line and passes through the origin shows that:

Rate $\propto$ [A]

or Rate = k[A] (where k is the gradient of the graph)

HINT (*You may find it useful to look back to p. 44 if you have forgotten this.*)

KEY FACT *When the rate of a reaction involving a species A is proportional to [A] we say that the reaction is **first order** with respect to A.*

The table below shows (idealised!) results from a second experiment studying a different reaction. The effect of changing the concentration of a reagent B on rate of reaction is shown. *Fig. 10* shows a graph of initial concentration of B against initial rate.

The above graph of rate against concentration is obviously not a straight line, so that rate is *not* proportional to the concentration of B; More importantly, the fact that the graph is not a straight

Initial [B]/ mol dm^{-3}	Initial rate of reaction/ mol dm^{-3} s^{-1}
0.100	0.500
0.200	2.00
0.300	4.50
0.400	8.00
0.500	12.5
0.600	18.0
0.700	24.5
0.800	32.0
0.900	40.5

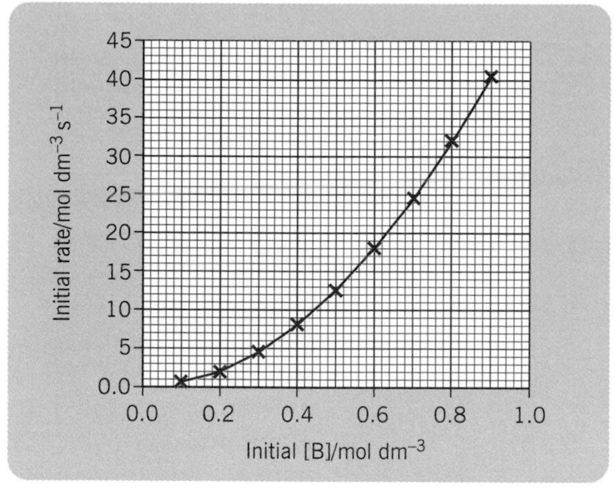

Fig. 10 *Rate of reaction against concentration of B.*

line means that we can draw no conclusion about the relationship between rate and [B] (other than they are not directly proportional). The reaction is not first order with respect to B.

To find out how they are related, you need to try drawing other graphs to find one which produces a straight line. In this case it helps to know some chemistry! If you don't, it is unlikely that you will produce a straight-line graph.

Another *possible* relationship between rate and concentration is

$$\text{Rate} = k[\text{B}]^2$$

If this relationship is correct, then a graph of rate against $[\text{B}]^2$ should be a straight line.

You have already seen (pp. 44–5) that straight-line graphs have the form:

$y = mx + c$ where m is the gradient and c the y-intercept.

$\text{Rate} = k\ [\text{B}]^2$

$y \quad m \ x$

To plot the graph to check whether rate and $[\text{B}]^2$ are related as shown, you need to calculate $[\text{B}]^2$; this has been done in the table below, and the graph plotted (*Fig. 11*).

Initial [B]/ mol dm^{-3}	Initial [B]^2/ mol^2 dm^{-6}	Initial rate of reaction/ mol dm^{-3} s^{-1}
0.100	0.0100	0.500
0.200	0.0400	2.00
0.300	0.0900	4.50
0.400	0.160	8.00
0.500	0.250	12.5
0.600	0.360	18.0
0.700	0.490	24.5
0.800	0.640	32.0
0.900	0.810	40.5

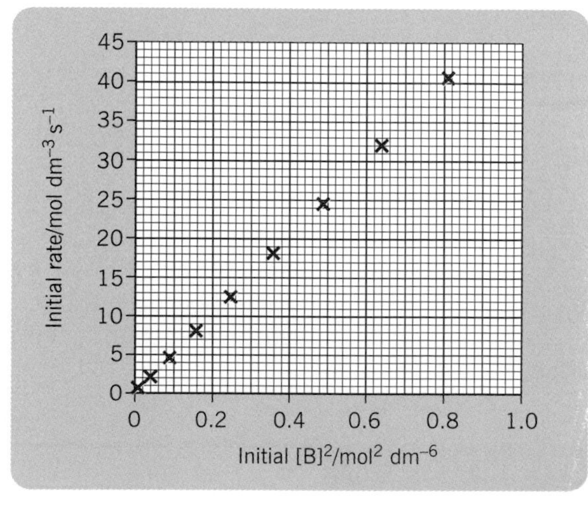

Fig. 11 *Rate of reaction against [B]^2*

Question: What can you deduce from *Fig. 11*?

Answer: This graph is a straight line and passes through the origin, so that we now know that the relationship between rate and [B] is:

$$\text{Rate} = k[B]^2$$

7.5 Spectra as graphs

Infra-red

Infra-red spectra plot information about the extent of absorption of infra-red radiation at different wavelengths. To make the numbers more manageable, the x-axis usually displays **wave number** (which is $\frac{1}{\text{wavelength}}$ in cm^{-1}) rather than actual wavelength. *Fig. 12* shows the infra-red spectrum of ethanol:

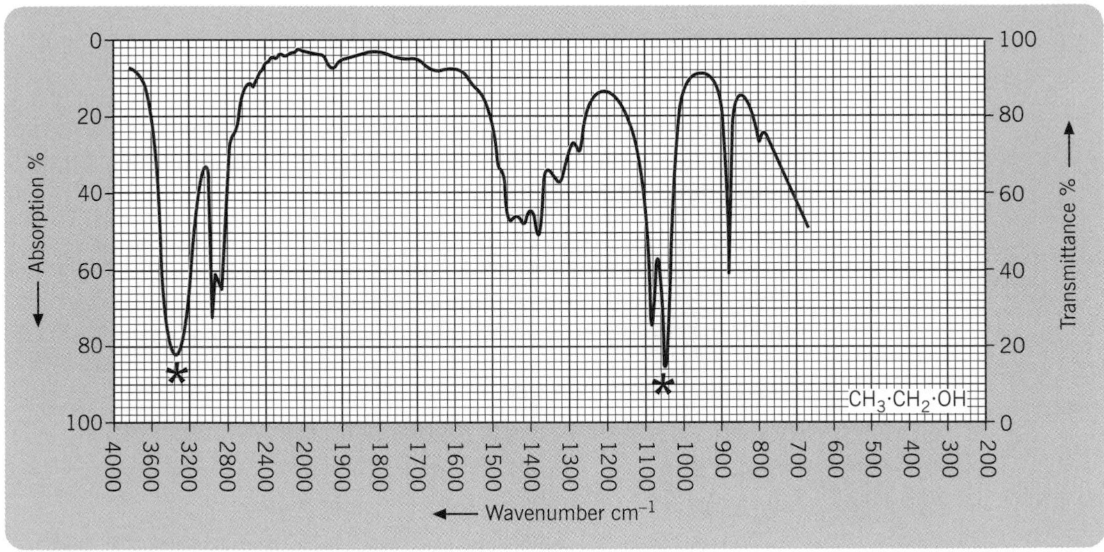

Fig. 12 *IR spectrum of ethanol*

Different functional groups in organic molecules absorb infra-red radiation of different wave numbers. By measuring the wave number of a peak from the spectrum and consulting tables of values, it is possible to decide which **functional groups** are —OH groups.

You need to be careful when reading off wavelength, as the **scale** on the x-axis sometimes changes at 2000 cm^{-1}. Apart from that, all that you need to do is to take two printed scale figures and estimate by eye the value between these that the absorption occurs, but be careful because the scale is not linear. As the same functional group in different molecules absorbs at slightly different wave numbers, estimating by eye is sufficiently accurate.

Try It Yourself

Exercise 7.5.1

1 Estimate the wave number of the absorptions marked with asterisks (*).

The height (or depth) of a peak enables you to read the percentage absorption, but this is not usually necessary. Most tables classify absorptions as strong (*s*), medium (*m*) or weak (*w*), and the actual percentage is not important.

Ultra-violet

Ultra-violet spectra give similar information in the ultra-violet region of the spectrum. This gives some information about unsaturated parts of the molecule (i.e. those regions containing double bonds). This is probably the least important type of spectrum and will not be discussed further here.

Nuclear magnetic resonance

NMR spectra display information about the frequencies of radiation (in the radio part of the spectrum) absorbed by hydrogen (and some other) nuclei when placed in a strong magnetic field. The interpretation of these is similar to that of infra-red spectra. You need to read the value of absorption from the *x*-axis (in this case, chemical shift measured in parts per million, p.p.m.) and consult a data table to discover the type of grouping present in the molecule.

The peak height (or more accurately, area) is proportional to the number of hydrogen atoms present, but this is always measured electronically.

Fig. 13 shows the low resolution spectrum of ethanol:

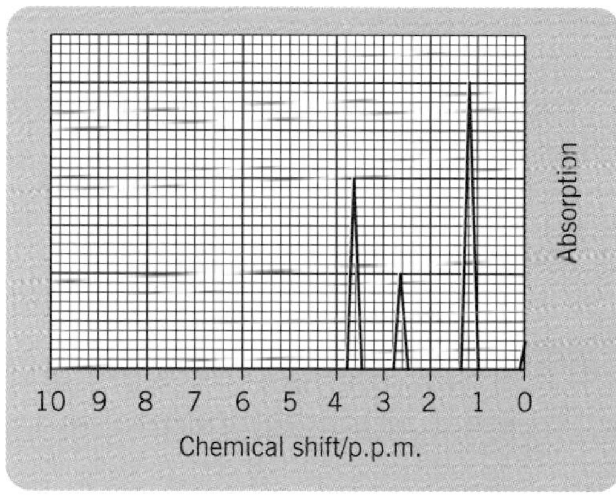

Fig. 13 *NMR of ethanol.*

Try It Yourself

Exercise 7.5.2

1 Estimate the chemical shift of each of the absorptions in the ethanol spectrum (*Fig. 13*).

The NMR spectra we have considered here are measured at **low resolution**. It is more usual to measure them at **high resolution**, which shows the finer detail in the peaks. This gives additional structural information (specifically about hydrogen atoms on *neighbouring* carbon atoms). If you need to know about this, the information can be found in A-level textbooks. It will not be considered further here.

Atomic emission spectra

Atomic emission spectra are those produced by applying a high voltage to a gas at low pressure. The one most commonly discussed is the hydrogen spectrum, as this led to the first suggestion that electrons in atoms could possess only certain energies – the existence of energy levels.

This is not strictly speaking a graph, as nothing is plotted vertically. The spectrum simply shows the wavelengths (and colours) of light emitted (*Fig. 14*). The wavelength can be measured in the way already described for infra-red spectra above.

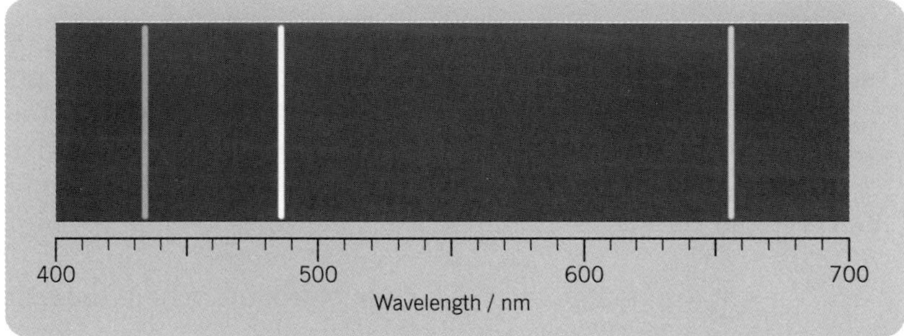

Fig. 14 *Emission spectrum of hydrogen.*

Try It Yourself — **Exercise 7.5.3**

1 Estimate the wavelengths of the lines in the hydrogen spectrum (*Fig. 14*). Convert the values you obtain to frequencies.

Mass spectra

The **mass spectrum** of an element shows the mass numbers and relative amounts (**relative abundances**) of the isotopes present in an element. The relative abundance is shown by the *height* of each peak, and can be read off the scale on the *y*-axis (*Fig. 15*).

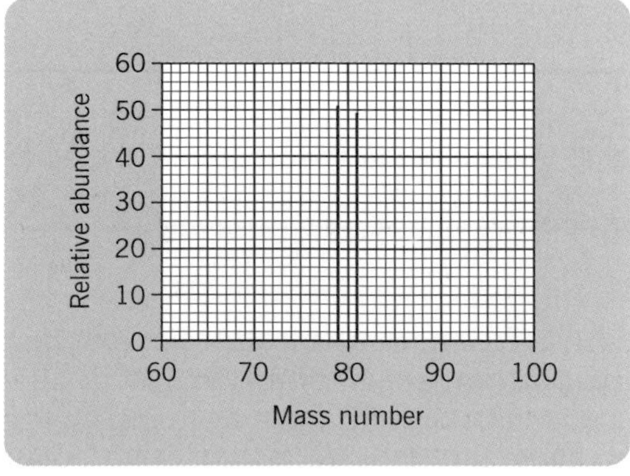

Fig. 15 *Mass spectrum of bromine atoms.*

This shows that naturally occurring bromine consists of two isotopes, ^{79}Br and ^{81}Br, present in approximately equal proportions (50.5% and 49.5% respectively).

This enables you to calculate the relative atomic mass, that is the **weighted mean** relative mass of the atoms. This is done as follows:

In 1000 bromine atoms there will be 505 ^{79}Br and 495 ^{81}Br.

Mass of 505 ^{79}Br atoms	$= 505 \times 79$	= 39 895 a.m.u.
Mass of 495 ^{81}Br atoms	$= 495 \times 81$	= 40 095 a.m.u.
Total mass of 1000 atoms	= 39 895 + 40 095	= 79 990 a.m.u.
Mean mass of 1 atom	$= \dfrac{79\,990}{1000}$	= 79.99 (80) a.m.u.

If you look in a data book (or at a Periodic Table), you will find the relative atomic mass of bromine given as 80.

Try It Yourself

Exercise 7.5.4

1 *Fig. 16* shows the mass spectrum of a sample of naturally occurring chlorine atoms.

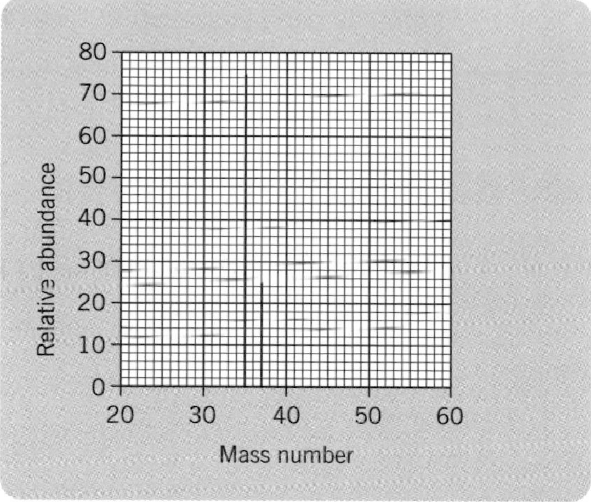

Fig. 16 *Mass spectrum of chlorine atoms.*

(a) Measure the mass numbers and relative abundance of each isotope.

(b) Use the figures from (part a) to calculate the (weighted mean) relative atomic mass of chlorine.

Mass spectroscopy is also used to measure the relative molecular masses of organic compounds. These spectra tend to be more complex than those we have looked at for elements. Organic molecules tend to disintegrate in the mass spectrometer, so that peaks are seen for the various fragments, as well as for the whole molecule. The relative molecular mass is obtained from the peak appearing at the highest mass. The masses of some of the other peaks can be used to identify the fragments into which the original molecule disintegrates. If these fragments can be fitted together, it is often possible to work out the structure of the original molecule.

Try It Yourself

Exercise 7.5.5

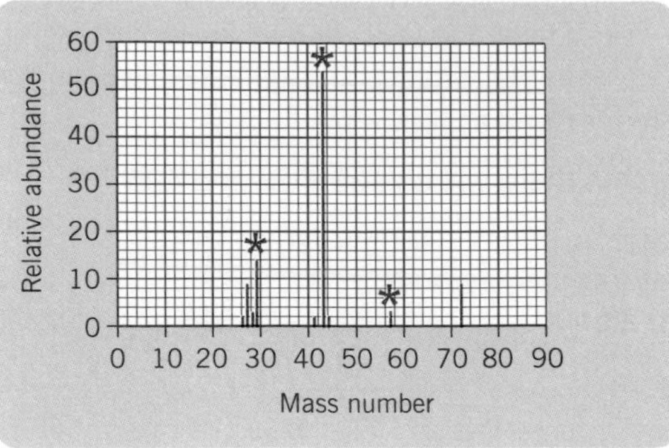

Fig. 17 *Mass spectrum of an organic compound.*

1. Use the mass spectrum in *Fig, 17* to work out the relative molecular mass of the compound.
2. Work out the masses of the fragments which cause the peaks marked with asterisks.
3. Suggest formulae for the fragments.
4. Suggest a way in which the fragments can fit together to give the original molecule.

7.6 Frequency distribution curves

For simple alkanes (saturated hydrocarbons), there is only one possible structure for each molecular formula. However, for those with four or more carbon atoms, more than one structure is possible. Table 1 shows the number of different structures that exist for alkanes containing up to 12 carbon atoms.

Table 1

No. of carbon atoms in molecule	No. of isomers
1	1
2	1
3	1
4	2
5	3
6	7
7	9
8	18
9	35
10	75
11	159
12	355

Fig. 18 shows the information displayed as a bar chart.
The data is *discontinuous* and is best displayed as shown as a bar chart. *Fig. 19* shows the same information displayed as a line graph. This is *not* an appropriate form; there is

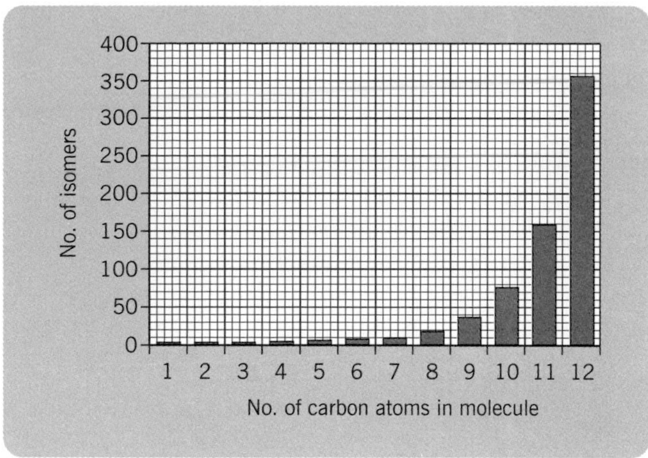

Fig. 18 *Isomers of alkanes.*

obviously no significance in the number of isomers apparently possible for an alkane with 7.5 carbon atoms, for example.

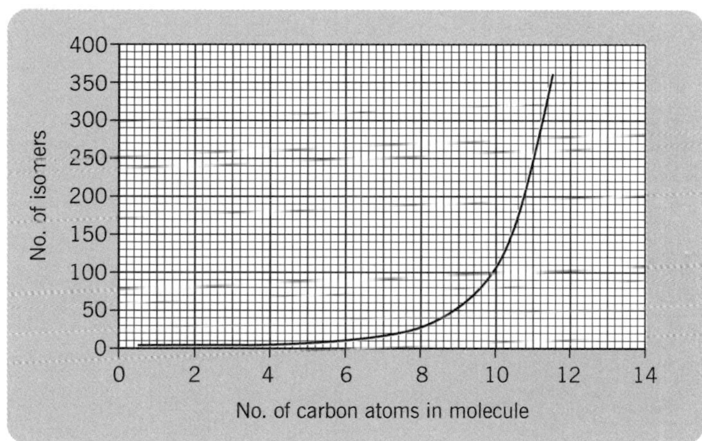

Fig. 19* *Incorrect *use of line graph.*

Fig. 20 shows a type of graph known as a **frequency distribution curve**. This is appropriate for *continuous* data, such as the fraction of molecules in a gas that have a particular energy.

Question: What can you learn from this curve?

Answer:

- There are no molecules with zero energy.
- There are few molecules with very low energies.
- The number of molecules with a particular energy increases to a maximum and then decreases again.

Question: How would you describe the energy at which the peak of this graph occurs?

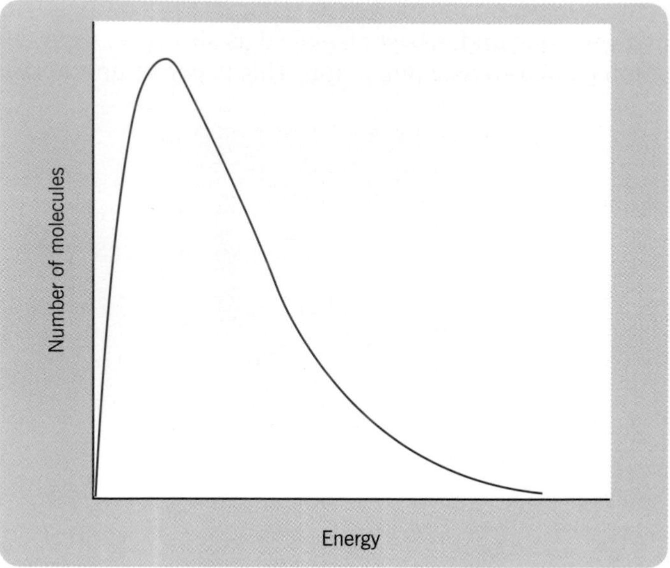

Fig. 20 *Number of molecules with a given energy.*

Answer: It is the **most probable energy** – i.e. the energy that the molecules are most likely to have.

At higher temperatures, the distribution of energies changes as shown in *Fig. 21*.

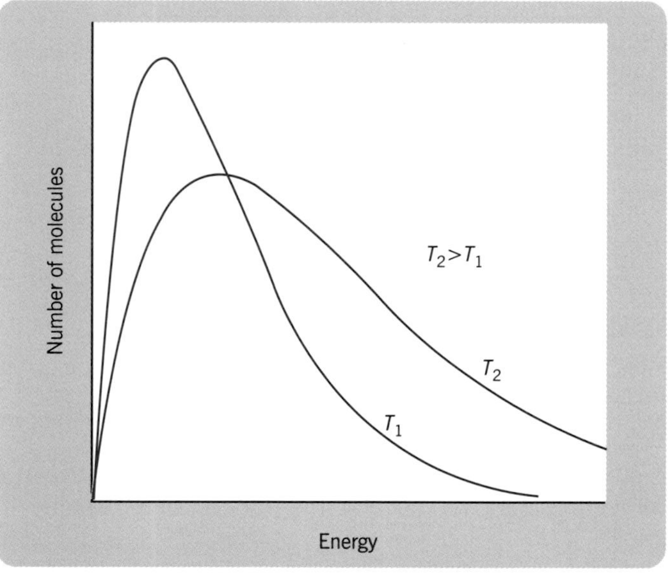

Fig. 21 *Number of molecules at different temperatures.*

You should note that:

- the general shape of the curve is the same
- the most probable energy is now higher
- the total area under the curve is the same in each case; it represents the total number of molecules, which is of course unchanged.

This curve is very useful in accounting for the rate of reaction under different conditions. In *Fig. 22*, E_a represents the **activation energy** for the reaction, i.e. the minimum energy that colliding molecules must possess if they are to react. Collisions involving molecules with energy lower than E_a do not result in a reaction. The coloured area to the right of E_a represents the fraction of the molecules with energy greater than or equal to E_a.

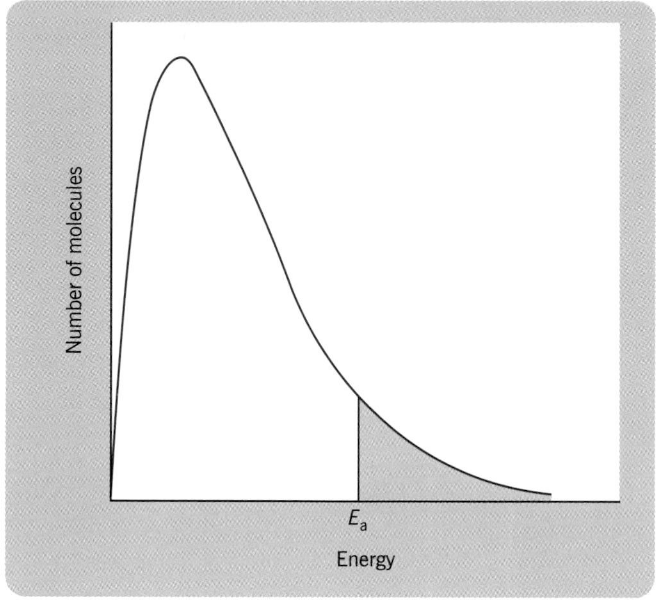

Fig. 22 *Activation energy,* E_a.

If E_a is marked on *Fig. 21*, it can be seen that the coloured area is much greater at the higher temperature (*Fig. 23*). Thus, when the temperature is increased, a greater fraction of collisions are successful; this is the major cause of increased rate of reaction at higher temperature.

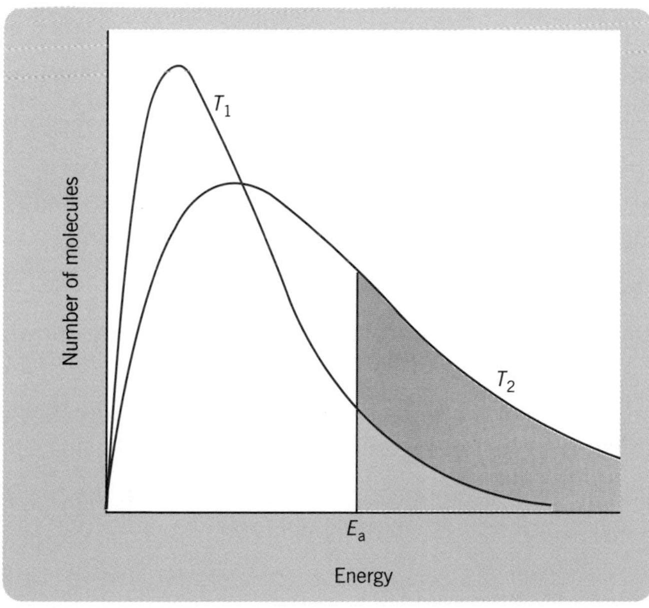

Fig. 23 *Number of molecules reacting at different temperatures.*

When a **catalyst** is added to a reaction, it frequently alters the reaction route. The new one has a lower activation energy. This increases the fraction of molecules able to react, and leads to more collisions being successful. The rate of the reaction is thus increased. This is shown in *Fig. 24*, where E_a represents the activation energy of the uncatalysed reaction; $E_{cat.}$ of the catalysed process.

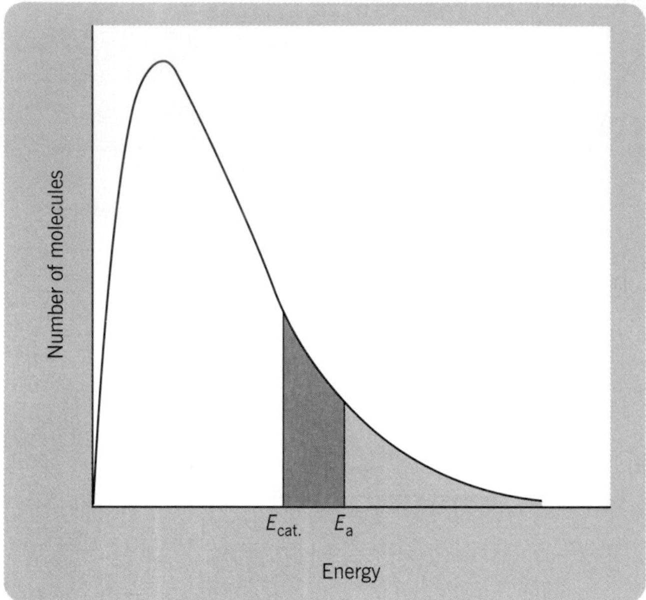

Fig. 24 *The effect of a catalyst on the number of molecules reacting.*

Summary

Graphs can be used to represent data.

Different types of graph are appropriate for different types of data e.g.

- line graphs for *continuous* data (e.g. the variation in concentration of a reactant with time)
- bar charts or histograms for *discontinuous* data (e.g. electronegativities of different elements)
- pie charts to show the relative importance of different components (e.g. the relative amount in each fraction obtained from a fractionating column).

Graphs can be used to obtain information e.g.

- from the gradient (giving rate of reaction for a graph of concentration of a reactant against time)
- a value between those measured by *interpolation* (e.g. percentage of ammonia in an equilibrium mixture at a pressure between two measured values)
- a value outside the measured range by *extrapolation* (e.g. the boiling point of eicosane, $C_{20}H_{42}$, from a graph of boiling point against relative molecular mass for the first 15 alkanes).
- You can find out how two variables are related by plotting a graph:

  The graph *must* be a straight line.

- If it is a straight line and passes through the *origin*, then whatever is plotted on the *y*-axis is related to whatever is plotted on the *x*-axis by $y = mx$ where m is the gradient.

For example:

- If a graph of distance (d) against time (t) is a straight line, then $d = kt$. This is true for something travelling at constant speed.
- If a graph of distance (d) against $(\text{time})^2$ is a straight line then $d = kt^2$. This is true for something falling under gravity.
- If a graph of initial concentration (c) against $\frac{1}{\text{time}}$ is a straight line, then $c = k.\frac{1}{t}$.

Spectra are a type of graph e.g.

- infra-red spectra show the wavenumbers of IR radiation absorbed, giving information about the *functional groups* present in a molecule
- atomic emission spectra show the wavelength/frequency of visible light absorbed or emitted by atoms, giving information about the differences between *electronic energy levels* in an atom
- nuclear magnetic resonance spectra show absorption of radio waves by nuclei in magnetic fields, giving information about the *chemical environment* of ^1H nuclei (amongst others) in a molecule
- mass spectra show the relative abundances of *isotopes* present in an element, leading to values for relative atomic mass
- mass spectra also show the mass of the molecular ion and important fragments of an organic molecule, providing information which helps us to work out its *structure*.

Frequency distribution curves can be used to show:

- the relative numbers of molecules having particular energies
- the fraction of particles having more than a threshold amount of energy in a liquid or a gas.

This helps to explain:

- the increase in rate of reaction with increased temperature
- the increase in rate of reaction when a catalyst is added.

Answers to Try It Yourself Questions

Exercise 7.2.1, *p. 50*

1 A line graph as the data is continuous (and gives a *straight* line).
2 Bar chart as the data is discontinuous – each bar will represent one hydrocarbon so that they are all the same width. This makes it a bar chart rather than a histogram.
3 A pie chart shows the relative proportions most clearly.

Exercise 7.3.1, *p. 52*

1 0.55 mol dm^{-3} s^{-1}
2 1.2 mol dm^{-3}

Exercise 7.3.2, *p. 52*

1 –9000 kJ mol^{-1}
2 –13400 kJ mol^{-1}

Exercise 7.5.1, *p. 56*

1 3360 cm^{-1}
1050 cm^{-1}

Exercise 7.5.2, *p. 57*

1 1.2 ppm
2.6 ppm
3.6 ppm

Exercise 7.5.3, *p. 58*

1 434 nm	486 nm	656 nm
6.91×10^{14} Hz	6.17×10^{14} Hz	4.57×10^{14} Hz

Exercise 7.5.4, *p. 59*

1 35.5

Exercise 7.5.5, *p. 60*

1 72
2 57; 43; 29
3 29 = CH_2CH_3; 43 = $—COCH_3$; 57 = $—COCH_2CH_3$
4 $CH_3COCH_2CH_3$

Chapter 8

Advanced graphs

After completing this chapter you should be able to:

- *find the value of* n *for data obeying the law* $a = b^n$ *by plotting a single graph*
- *find the values of* p *and* n *for data obeying the law* $a = p \times b^n$.

8.1 Logs and graphs

Chemists are often faced with a lot of experimental measurements from which they want to work out the relationship between two variables.

A common example is trying to find the relationship between the *rate* of a reaction and the *concentration* of one (or more) of the reagents. To do this, you need to measure the rate of reaction for several different concentrations of the reagent being investigated. How this is done depends on the particular reaction and will not be discussed here.

You might obtain results such as those shown in the table below.

[Reagent]/ mol dm^{-3}	Initial rate of reaction/ mol dm^{-3} s^{-1}
0.100	0.500
0.200	2.00
0.300	4.50
0.400	8.00
0.500	12.5
0.600	18.0
0.700	24.5
0.800	32.0
0.900	40.5

These are 'ideal' results. 'Real' results would be affected by experimental error and produce graphs on which the points are more scattered. This particular problem is considered in Chapter 9.

These results have already been studied in the previous chapter (see p. 55) where:

a graph of **rate** against **[reagent]** is *not* a straight line,
but a graph of **rate** against $\mathbf{[reagent]^2}$ is a straight line.

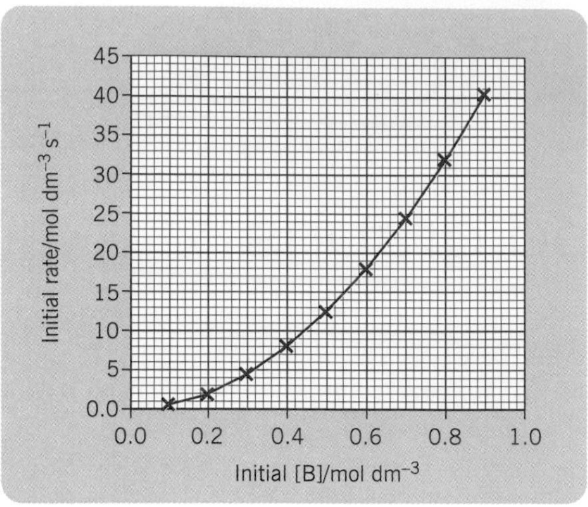

Fig. 1 *Rate of reaction against [B].*

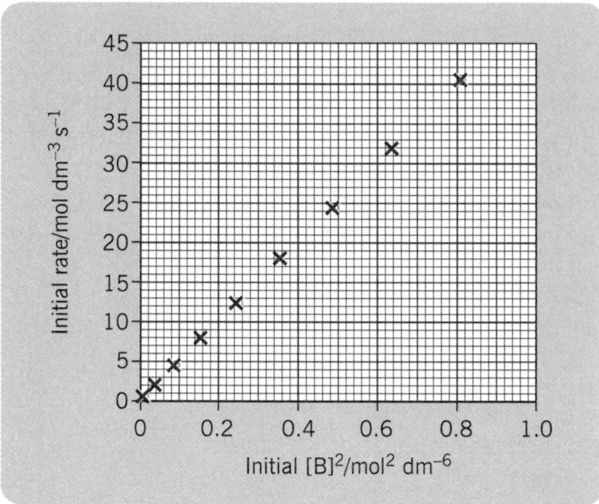

Fig. 2 *Rate of reaction against [B]^2.*

This shows that:

$$\text{Rate} = k[\text{B}]^2$$

HINT

If you do not understand this, it would be helpful to go back to p. 43 and read through the section again.

In chemistry, it is often possible to use this method of trial-and-error, plotting graphs with the concentration of the reagent raised to different powers; but if you do not plot the correct graph, you will not find the straight line!

We can find the relationship by plotting a single graph, involving **logs**.

Try It Yourself

Exercise 8.1.1

1 Use your calculator to complete the following table:

Number	Number in scientific notation	Log(number)
1		
10		
100		
1000		

Question: Look carefully at your answer in the 'number in scientific notation' and 'log(number)' columns above. Can you see a pattern?

Answer: You may notice $\mathbf{\log 10^x = x}$ Check it for yourself.

This is a general result – even for numbers where the relationship is less obvious.

KEY FACT $\log 10^x = x$
e.g. $\log 10^{2.4} = 2.4$

Again check it for yourself.

Before calculators were so readily available, logs were used for multiplication and division.

Try It Yourself

Exercise 8.1.2

2 Complete the following table.

x	$\log x$	y	$\log y$	$x \times y$	$\log(x \times y)$
10		10			
10		100			
100		100			
1000		1000			

Question: Can you see a pattern?
Answer: You may be able to see that: $\mathbf{\log(x \times y) = \log x + \log y}$ Check it for yourself.

KEY FACT $\log(x \times y) = \log x + \log y$

Try It Yourself

Exercise 8.1.3

1 By treating $\log x^2$ as $\log (x \times x)$, simplify $\log x^2$.

2 Simplify $\log x^3$.

3 From your answers to questions 1 and 2, suggest a simplified form for $\log x^n$.

Plotting log graphs

Suppose that the relationship between two variables a and b is:

$$a = b^n$$

Take logs of both sides: $\log a = n \log b$

The equation $\log a = n \log b$ will produce a straight-line graph if $\log a$ is plotted on the y-axis and $\log b$ is plotted on the x-axis. The graph will have a gradient of n.

$$\log a = n \log b$$
$$(y = mx)$$

This method produces a value for n from a single graph, whatever the value of n. It will work if n is positive or negative, or even if n is not a whole number.

KEY FACT *The more general form for this relationship is:*

$$a = p \times b^n$$

Taking logs of both sides gives:

$$\log a = \log p + n \log b$$

which we can write as:

$$\log a = n \log b + \log p$$

Question: What graph would you plot to find the values of n and p in the above equation?

Answer:

$$\log a = n \log b + \log p$$
$$(y = mx + c)$$

A graph of $\log a$ (y-axis) against $\log b$ (x-axis) will have a gradient of n and an intercept on the y-axis (see p. 44) of $\log p$.

Applying this method to the data from p. 67 gives the following results.

[Reagent]/ mol dm^{-3}	Initial rate of reaction/ mol dm^{-3} s^{-1}	Log[reagent]	Log(initial rate)
0.100	0.500	−1.000	−0.301
0.200	2.00	−0.699	0.301
0.300	4.50	−0.523	0.653
0.400	8.00	−0.398	0.903
0.500	12.5	−0.301	1.097
0.600	18.0	−0.222	1.255
0.700	24.5	−0.155	1.389
0.800	32.0	−0.097	1.505
0.900	40.5	−0.046	1.607

HINT *Note that quantities involving logs do not have any units.*

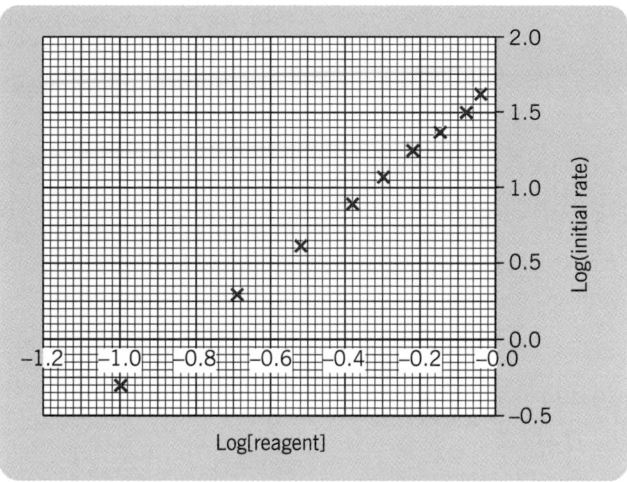

Fig. 3 *Log graph of [reagent] and initial rate.*

The gradient of this graph is +2 (you may care to check this).

However, if you look at the graph again, you will see that it does not pass through the origin. The general equation for a straight line is:

$y = mx + c$ where c is the intercept on the y-axis.

In this case, the intercept is 1.7, so that the equation of the line is:

$\log(\text{rate}) = 2 \times \log[\text{reagent}] + 1.7$

or $\text{rate} = k\,[\text{reagent}]^2$

where $\log k = 1.7$

i.e. $k = 50.1$

So the final form of the equation is:

$\text{Rate} = 50.1 \times [\text{reagent}]^2$

The Arrhenius equation

One other example using a log graph that you may meet involves the **Arrhenius equation**:

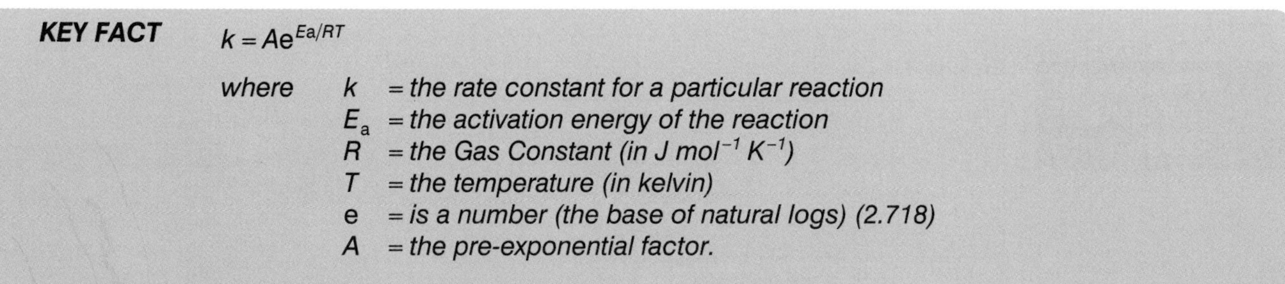

KEY FACT $k = Ae^{Ea/RT}$

where k = *the rate constant for a particular reaction*
E_a = *the activation energy of the reaction*
R = *the Gas Constant (in J mol*$^{-1}$ *K*$^{-1}$*)*
T = *the temperature (in kelvin)*
e = *is a number (the base of natural logs) (2.718)*
A = *the pre-exponential factor.*

HINT

This equation models the idea that reactions proceed through particles colliding with one another. They must have sufficient energy to react (the activation energy, E_a). The factor $e^{-E_a/RT}$ represents the probability that a molecule will possess this. A reflects the requirement that they need to collide at the correct orientation – for example, with the 'reactive bits' next to one another. It is unlikely that you need to know this for your examination!

Taking logs of both sides of this equation:

$$\log k = \log A - E_a/RT \times \log e$$

The value of log e is 0.434, so the equation can be rearranged as follows:

$$\log k = -\frac{0.434}{RT} + \log A$$

$$\log k = -\left(\frac{0.434}{R} \times \frac{1}{T}\right) + \log A$$

$$(y = mx + c)$$

A graph of log k (y-axis) against $\frac{1}{T}$ (x-axis) will be a straight line, with:

$$\text{Gradient} = \frac{-0.434E_a}{R}$$

$$E_a = \frac{-\text{gradient} \cdot R}{0.434}$$

So, if the rate constant of a reaction is measured at different temperatures, and a graph of log k against $\frac{1}{T}$ is plotted, the activation energy can be found from the gradient.

There are several standard A-level experiments in which this is done; you may carry out one of them.

Summary

You should now:

- know that for the equation $a = b^n$ a graph of log a (y-axis) against log b (x-axis) will be a straight line of gradient n
- know that for the equation $a = pb^n$ a graph of log a (y-axis) against log b (x-axis) will have a y-intercept of log p.

Answers to Try It Yourself Questions

Exercise 8.1.1, *p. 69*

Number	Number in scientific notation	Log(number)
1	1×10^0	0
10	1×10^1	1
100	1×10^2	2
1000	1×10^3	3

Exercise 8.1.2, *p. 69*

x	log x	y	log y	$x \times y$	log $(x \times y)$
10	1	10	1	100 (10^2)	2
10	1	100	2	1000 (10^3)	3
100	2	100	2	10 000 (10^4)	4
1000	3	1000	3	1 000 000 (10^6)	6

Exercise 8.1.3, *p. 70*

1 2 log x
2 3 log x
3 n log x

Chapter 9

Graphs from experiments

After completing this chapter you should be able to:

- *draw a line which fits the results of your own experiment*
- *judge by eye which line is the best fit to your points*
- *calculate the 'line of best fit' for your data points*
- *draw the best curve which fits your points*
- *recognise unreliable data points which should be ignored.*

9.1. Introduction

The examples given in Chapters 6–8 have all been 'ideal' results, that is those which an 'ideal' (or perfect!) chemist working with 'ideal' (or perfect) equipment might have obtained.

In practice, neither ideal chemists nor ideal apparatus exist. The chemist has to use his or her judgement to decide whether the end-point has been reached, to take a reading after 30 s (not 29 s, not 31 s), to estimate whether the temperature is 34.5 °C or 34.7 °C (i.e. to estimate between the graduation marks on the thermometer).

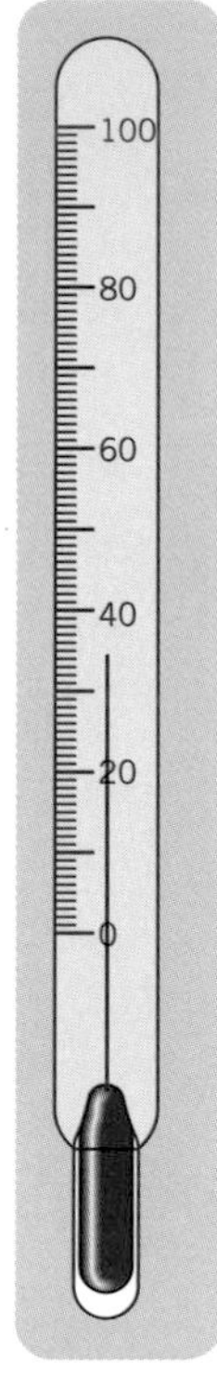

Fig. 1 *A lab thermometer.*

9.2 Limits of accuracy

There are **limits** on the accuracy with which any measuring instrument (balance, thermometer, pipette, burette, measuring cylinder) is manufactured and calibrated. The burette you use in your practical work is probably Grade B, so that the manufacturer guarantees that it is accurate to within ±0.1 cm^3. This means that if your end-point occurs at 30.0 cm^3 when you have approached it drop-wise, with every possible precaution to ensure accuracy, it is still somewhere between 29.9 cm^3 and 30.1 cm^3.

Or, to put it another way, if the 'correct' answer is 30.0 cm^3, however carefully and skilfully you use the apparatus, you are as likely to obtain a reading of 29.9 cm^3 as 30.0 cm^3.

Ideal vs. real results

An 'ideal' set of results from an experiment investigating how the rate of a reaction varies with the concentration of reagent A is shown at the top of page 75:

[A]/ mol dm^{-3}	Rate of reaction/ mol dm^{-3} s^{-1}
0.150	0.166
0.334	0.259
0.480	0.333
0.670	0.429
0.815	0.502
1.00	0.596
1.33	0.763
1.67	0.935

These results lead to the graph of rate against concentration as shown below (*Fig. 2*).

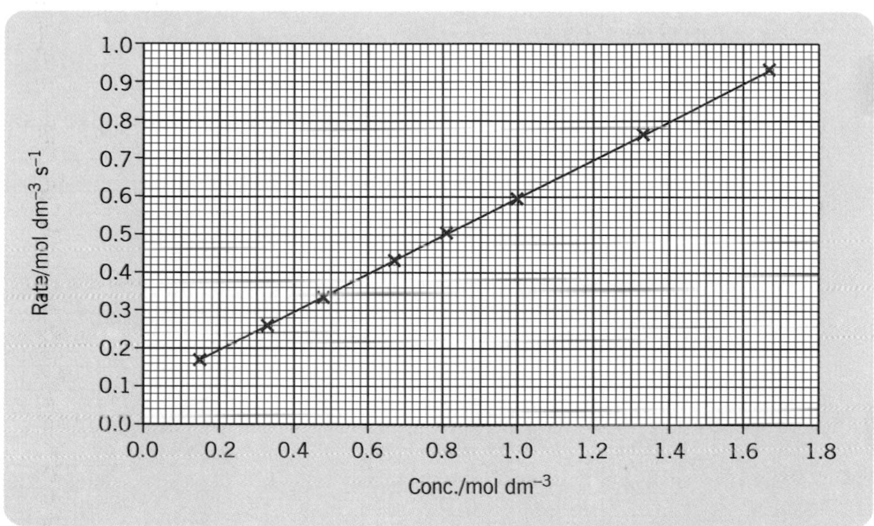

Fig. 2 *Rate of reaction against [A].*

However, the difficulties of making the measurements in this experiment, coupled with the errors caused by the manufacture and calibration of the apparatus are likely to lead to results more similar to those shown below (*Fig. 3*).

'Line of best fit'

Example

Claire obtained the following data for hydrogen peroxide:

[A]/mol dm^{-3}	Rate of reaction/mol dm^{-3} s^{-1}
0.150	0.201
0.334	0.272
0.480	0.372
0.670	0.385
0.815	0.554
1.00	0.583
1.33	0.786
1.67	0.929

These results produce the points on a graph as shown below (*Fig. 3*).

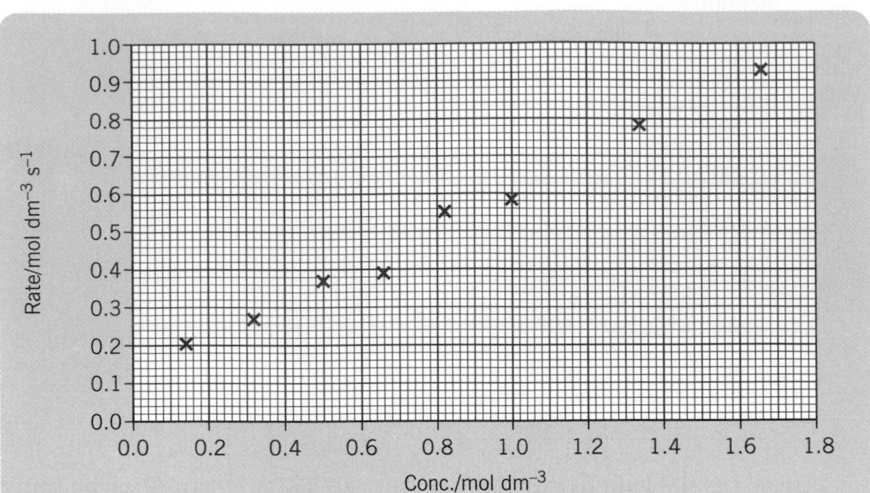

Fig. 3 *Rate of reaction against [A].*

Sometimes, you will only have to decide whether the points give a straight line or a curve. In that case, you have to try drawing a straight line and see if the points lie reasonably close to it. Ideally, the points that are not actually on the line should be distributed evenly on either side.

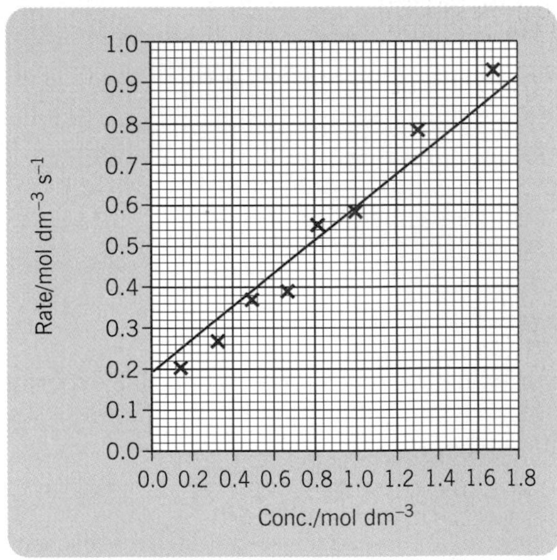

Fig. 4 *Incorrect line.*

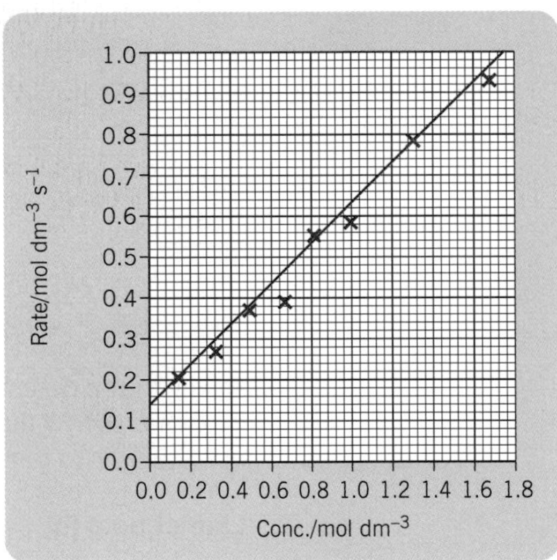

Fig. 5 *Incorrect line.*

Question: Neither show the 'line of best fit'; suggest what is wrong with each of them.

Answer: *Fig. 4* shows more points below than above the drawn line. Although the line for *Fig. 5* passes through several points, there are more points below than above. To draw the 'line of best fit', you just have to take a ruler lay it on the graph to find the best line, which passes as close as possible to the points (in your opinion), draw it in, and then decide whether or not your line looks reasonable.

Fig. 6 below shows the line of best fit for these points.

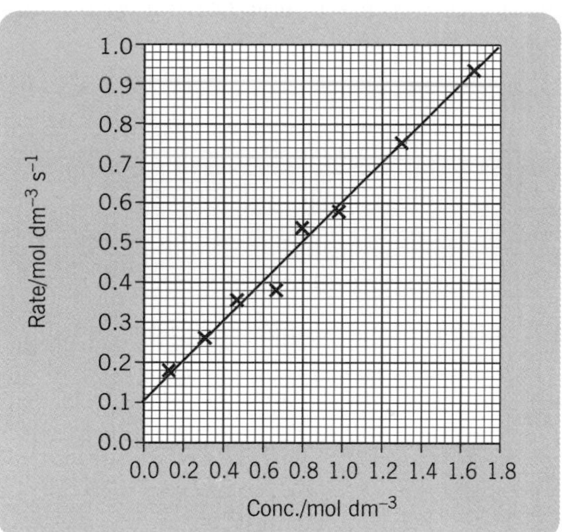

Fig. 6 *'Line of best fit'.*

If you need to measure the gradient of the line, do it in the usual way.
In this case, the gradient is 0.489.

HINT

You are unlikely to handle large amounts of data in your course. It is likely that the most appropriate approach is therefore to plot your graph by hand and draw what you consider the best straight line ('line of best fit').

Try It Yourself

Exercise 9.2.1

1 *Fig. 7* shows a scatter graph for the temperature of a solution as portions of aqueous sodium hydroxide are added to some sulphuric acid (the neutralisation reaction is exothermic). Work out the 'line of best fit' and estimate the intercept on the *y*-axis.

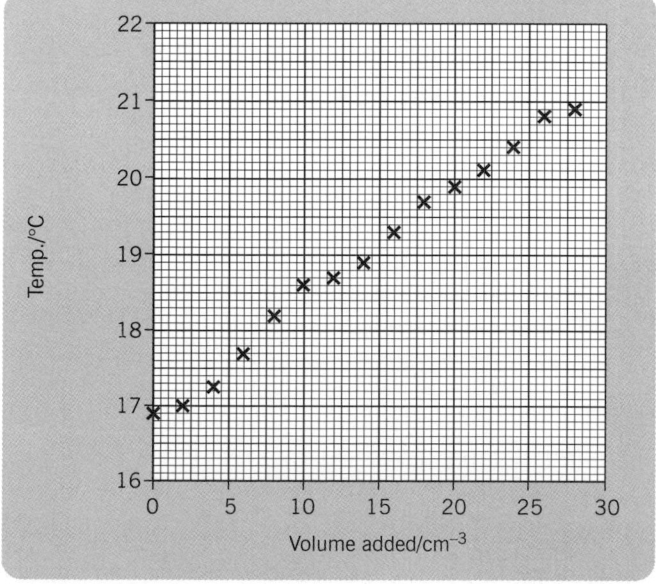

Fig. 7 *Neutralisation of H_2SO_4.*

2 *Fig. 8* shows a scatter graph for the concentration of aqueous iodine in equilibrium with different concentrations of iodine in an organic solvent. Work out the 'line of best fit' and estimate the intercept on the *y*-axis.

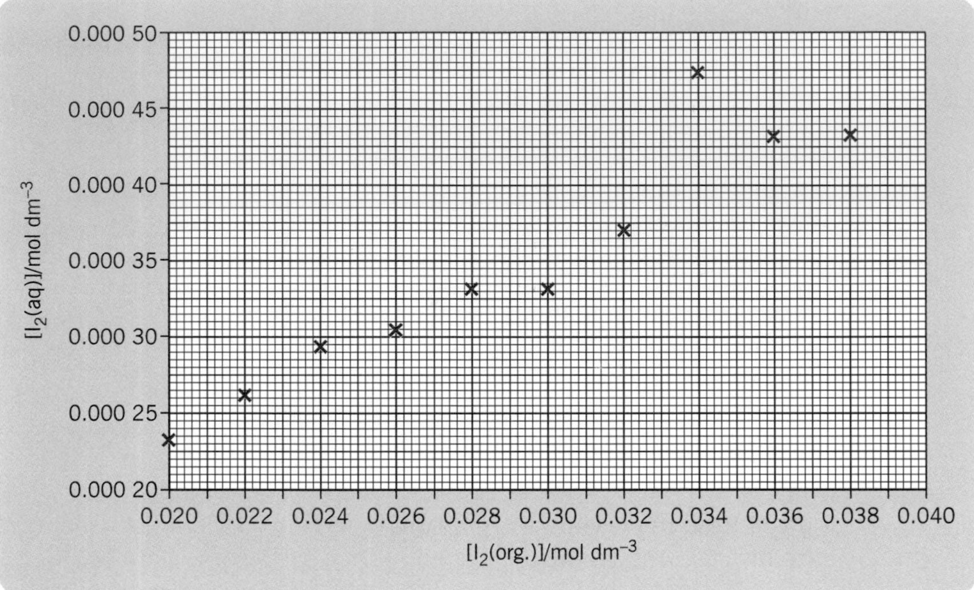

Fig. 8 *Equilibrium of I_2(aq.) and I_2(org.).*

9.3 The regression line

If you are handling large quantities of data, and *know that the graph should be a straight line*, you can determine the gradient and intercept by calculating the **regression line**. This calculation is a mathematical device which gives the line for which the total of the distances of all the points from the line is a minimum (i.e. the line passes as close as possible to all the points).

The process involves a fair amount of 'number crunching', calculating the values needed for substitution into the following equations:

KEY FACT *The 'line of best fit' will be:*

$$y = mx + c$$

where

$$m = \frac{\Sigma xy - (\Sigma x \Sigma y)}{N}$$

$$= \Sigma x^2 - \frac{(\Sigma x)}{N}$$

and

$$c = \frac{\Sigma y - m\Sigma x}{N}$$

where Σx = the sum of all the x values, Σy = the sum of all the y values.

Σxy = the sum of all the products of x and y, N = the number of readings taken.

HINT

You are unlikely to need to do this unless, as part of your course, you complete a project involving a large amount of data. If you do, you will probably find that it is available on your calculator – there is an example, in the Appendix at the end of this chapter, p. 83. It is also part of a statistics package on a computer.

You should be aware of one risk of using this technique. It will tell you the gradient and intercept of the 'line of best fit'. It will not make clear whether you should draw a straight line through the points; it will work equally well with data that does give a reasonably straight line and with data that produces a smooth curve. You really need to draw the graph to decide whether you are justified in drawing a straight line!

More sophisticated packages are available, which enable you to work out the best curve that fits data points. For example, the Casio range of calculators will calculate the values of a, b and c in the equation:

$$y = ax^2 + bx + c$$

which best fits your data points – but again using the method *assumes* that your points lie on a curve of this form. Even more sophisticated programs will produce curves with more complex equations that give better fits to your points, but making the same assumption. These methods need to be used cautiously – but they are beyond the scope of your current chemistry course.

9.4 Data that gives curves

You might carry out an experiment that results in a curve rather than a straight line. Again, the limitations of the apparatus will probably lead to the points lying either side of a curve rather than directly on it. In these circumstances, you must draw the best possible *smooth* curve which passes as close to all the points as possible (again, you are aiming to reduce the distance of all points from the curve to a minimum).

HINT

This is a matter of your judgement; there is no infallible method of doing this. The usual technique is to have a few 'dummy runs' with a pencil just above the paper, and then (after taking a deep breath!) drawing a single, smooth curve. Do not simply join up the points ('dot-to-dot'), and avoid the temptation to draw the curve in short sections so that the joins show.

Example

An example of an experiment that produces a curve is one which investigates the decomposition of hydrogen peroxide. A common way of measuring the rate of this reaction is to measure the volume of oxygen collected at regular time intervals. A set of 'ideal' results and the graph they produce are shown below (*Fig. 9*).

Time/s	Volume of O_2/cm^3
0	0.0
10	1.5
20	3.5
30	9.0
40	15.5
50	21.5
60	27.0
70	32.8
80	38.0
90	42.8
100	47.5
110	51.0
120	54.8
130	57.5

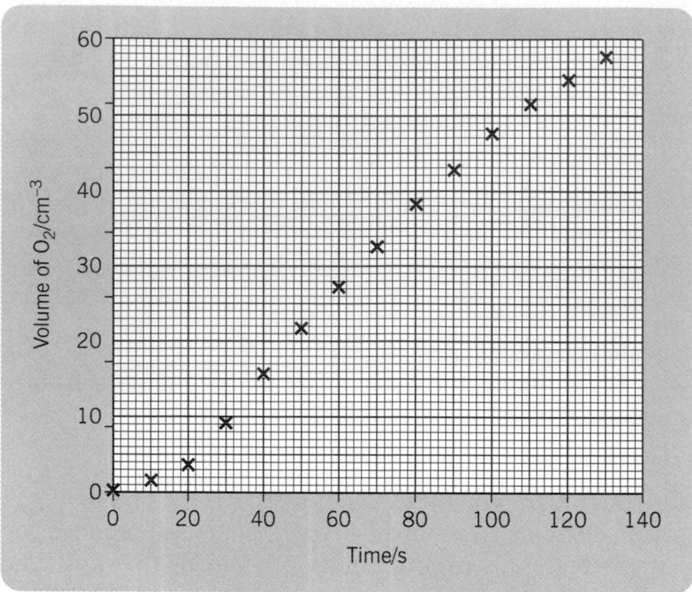

Fig. 9 *Decomposition of hydrogen peroxide (ideal).*

In practice, a set of results such as those shown below is more likely to be obtained.

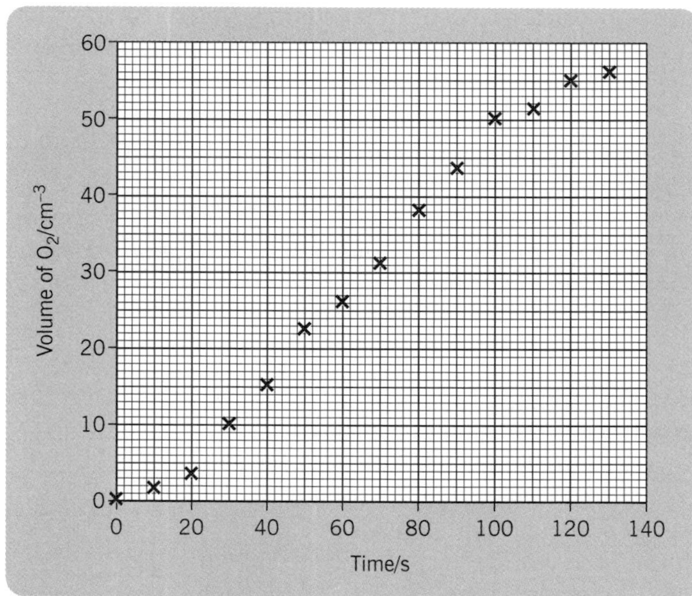

Fig. 10 *Decomposition of hydrogen peroxide (real)*

Time/s	Volume of O_2/ cm^3
0	0.0
10	1.5
20	3.5
30	10.1
40	15.0
50	22.5
60	26.0
70	31.0
80	38.0
90	43.5
100	49.8
110	51.0
120	54.8
130	56.1

These points should *not* simply be joined up (*Fig. 11*). Instead, a single *smooth curve* should be drawn as close to all the points as possible (*Fig. 12*).

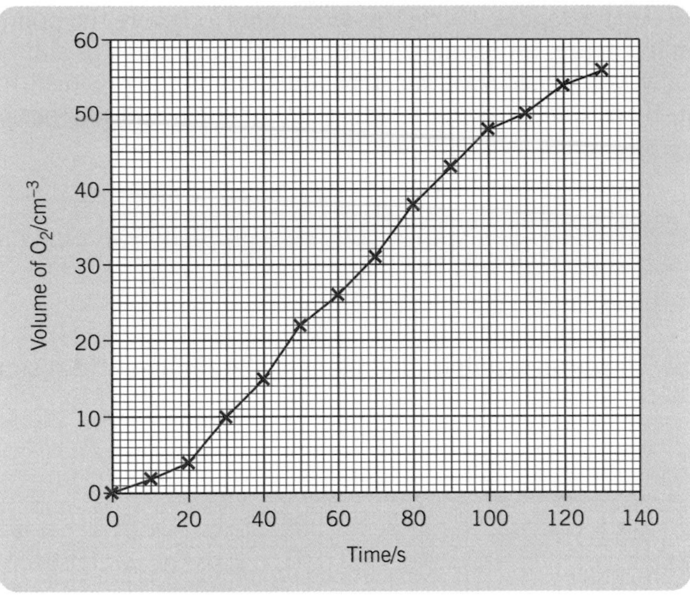

Fig. 11 Incorrect joining up.

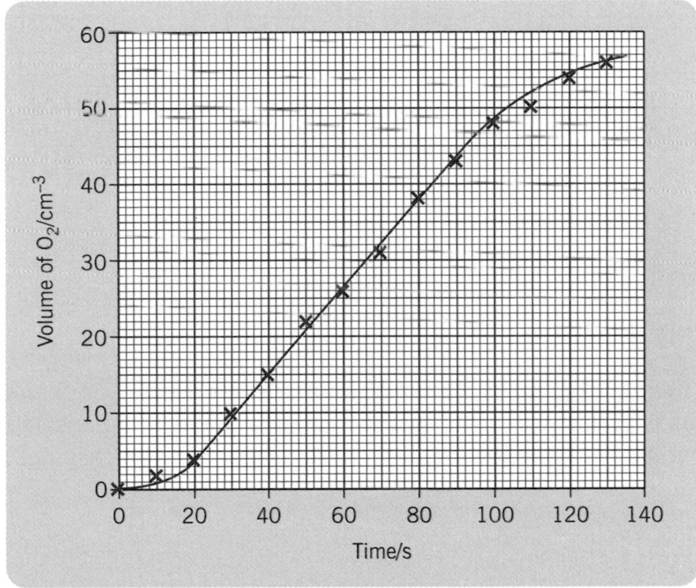

Fig. 12 Correct smooth curve.

9.5 Rogue points

Another problem that occurs with real experimental results is that of the **rogue point**. This is one that is so far away from the line or smooth curve that it cannot possibly be correct – you have either made a really serious mistake when taking the reading (e.g. missing a decimal point on a balance), or used the wrong reagent, or used insufficient reagent, etc.

Under these circumstances it is acceptable simply to ignore the point altogether (with some explanation if you are, for example, writing up the practical, or submitting a project). Of course, ideally, you would spot the error whilst doing the practical so that you can repeat the experiment, check that particular reading and obtain a correct one.

HINT *It is good practice to plot your results on a graph as you take them.*

Fig. 13 below shows an example of a rogue reading which is best simply ignored.

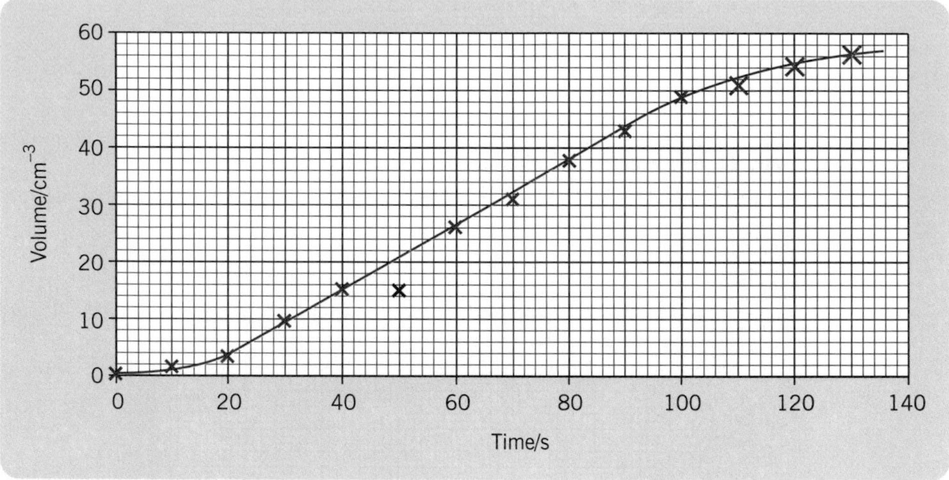

Fig. 13 *A rogue point.*

There is probably a fairly simple explanation for this event. For example, a common method of doing this experiment is to collect the gas in a graduated glass syringe. Occasionally these stick; this would lead to a low reading, as appears to be the case here.

Summary

- All experimental results have inevitable error.
- Drawing a graph lets you 'smooth out' some of the errors.
- Points may lie on a straight line or a curve; you can decide which it is by inspection.
- Draw a *straight* line or a *smooth* curve – don't just join the dots.
- You can draw the 'line of best fit' by inspection.
- If you have enough data you can use a regression line – but only if you are sure that the data should fit to a line rather than a curve.
- Rogue points can be ignored, provided that you add a note to your report.

Appendix

Using a Casio fx-83WA calculator to obtain the regression line

You first need to select the linear regression mode. To do this you press MODE

This produces the display:

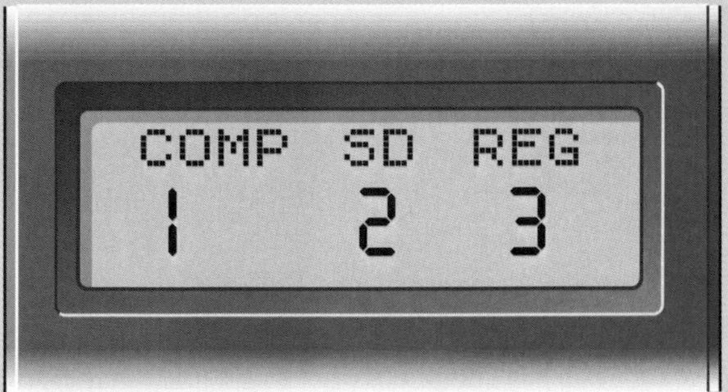

Choose 3, which produces the following display.

From this choose 1 (linear).

Clear all the data memories (as they will be used for storing and manipulating the data you are about to enter) using the following key sequence:

SHIFT Scl =

Here are Claire's results for the decomposition of hydrogen peroxide (p. 000):

[A]/ mol dm^{-3}	Rate of reaction/ mol dm^{-3} s^{-1}
0.150	0.201
0.334	0.272
0.480	0.372
0.670	0.385
0.815	0.554
1.00	0.583
1.33	0.786
1.67	0.929

These would be plotted with concentration on the x-axis and rate on the y-axis. Each pair of data points needs to be entered into the calculator in *that order*.

The general key sequence is:

[x_1] [,] [y_1] [DT] [x_2] [,] [y_2] [DT] etc.

where x_1 and y_1 represent the first values of [A] and rate respectively, from the above table.

The complete key sequence for the above data becomes:

[0.15] [,] [0.201] [DT]
[0.334] [,] [0.272] [DT]
[0.48] [,] [0.372] [DT]
[0.67] [,] [0.385] [DT]
[0.815] [,] [0.554] [DT]
[1] [,] [0.583] [DT]
[1.33] [,] [0.786] [DT]
[1.67] [,] [0.929] [DT]

For the regression line:

$$y = mx + c$$

The value of m is obtained from the following keys:

[SHIFT] [B] [=]

The value of c is obtained from the following keys:

[SHIFT] [A] [=]

If we apply this to the data shown above, the regression line is found to be:

$$y = 0.489x + 0.116$$

Exercises 9.2.1, *p. 77*

1 The calculated regression line (using a Casio *fx-83WA* calculator) has a gradient of 0.150 and an intercept on the y-axis of 16.9.

2 The calculated regression line has a gradient of 0.125 and an intercept on the y-axis of 0.000235. This value is the intercept on the y-axis of the graph *as printed*. Note that the y-axis is not at $x = 0$.

Exam Questions

Exam type questions to test understanding of Chapters 6–9

1 A 10.0 cm^3 sample of a solution containing 7.20 g dm^{-3} of a carboxylic acid was titrated against 0.0500 mol dm^{-3} sodium hydroxide. The pH readings shown in the table were taken.

Vol. NaOH/cm^3	0.0	2.5	5.0	7.5	10.0	14.0	15.0	16.0	17.5	20.0	22.5
pH	2.5	3.2	3.5	3.8	4.1	4.7	5.2	9.1	11.5	11.8	12.0

Use the graph (*Fig. 14*) plotted from these results to find the end-point of this titration. From this, calculate the concentration of the acid in mol dm^{-3}. You should assume that 1 mol of the acid reacts with 1 mol of sodium hydroxide.

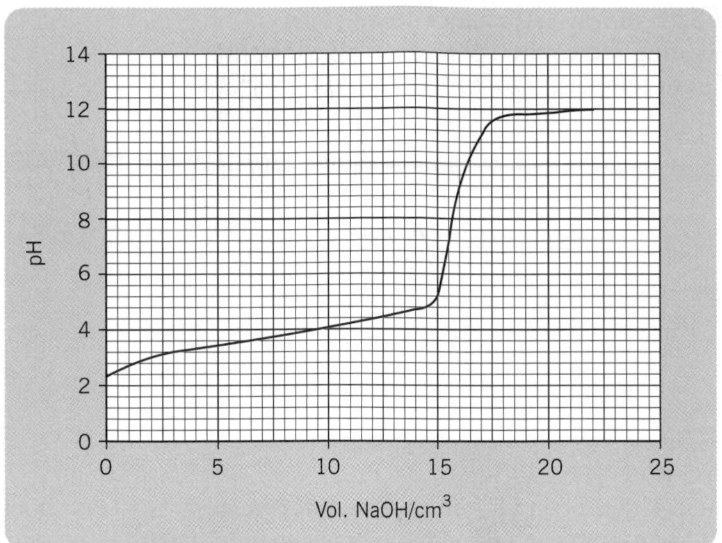

Fig. 14 Titration of carboxylic acid against sodium hydroxide.

2 *Fig. 15* below shows the percentage of ammonia in equilibrium mixtures at different temperatures and pressures:

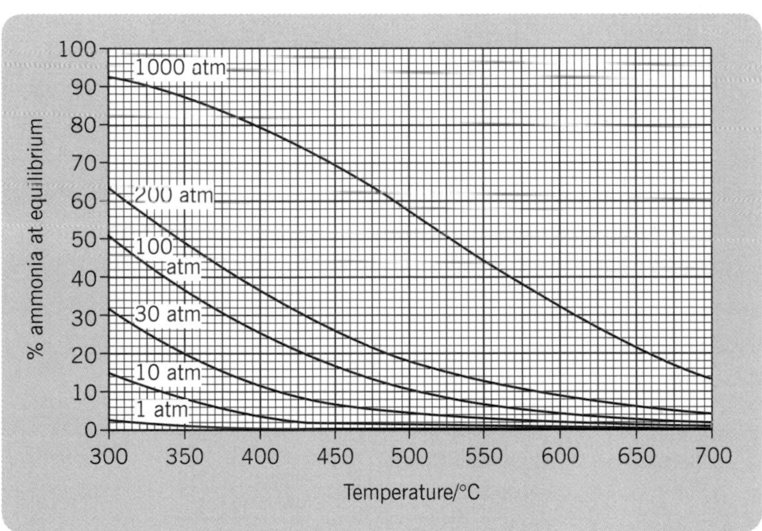

Fig. 15 Ammonia inequilibrium mixtures.

(a) Use the graphs to find the percentage of ammonia for the following sets of conditions:

(i) 30 atm and 300 °C.

(ii) 200 atm and 400 °C

(iii) 1000 atm and 500 °C

(b) Use the graphs to find the percentage of ammonia at 450 °C for each of the pressures shown. Use this to plot a graph of percentage of ammonia against pressure at 450 °C.

3 The data shown below were obtained by measuring the volume of a gas in a syringe at different temperatures but at constant pressure:

Temp./ K	Volume/ cm^3
275	25.0
285	25.9
295	26.8
305	27.7
315	28.6
325	29.6
335	30.5
345	31.4

The data have been plotted on three graphs (*Figs. 16–18*). Choose the one which enables you to work out the relationship between volume and temperature.

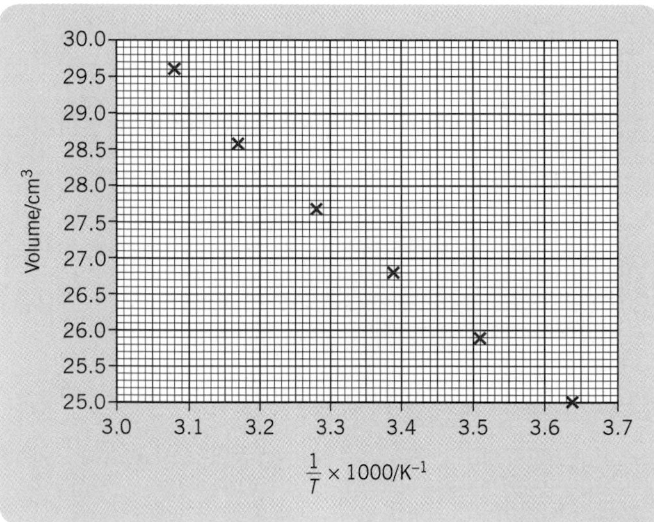

Fig. 16

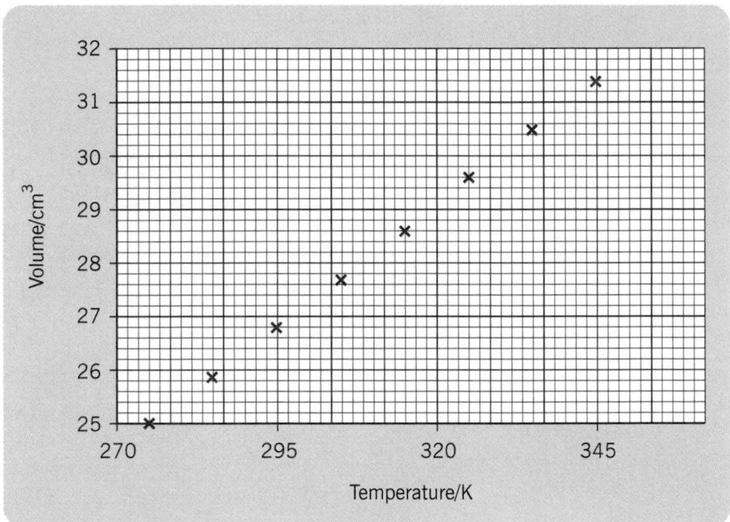

Fig. 17

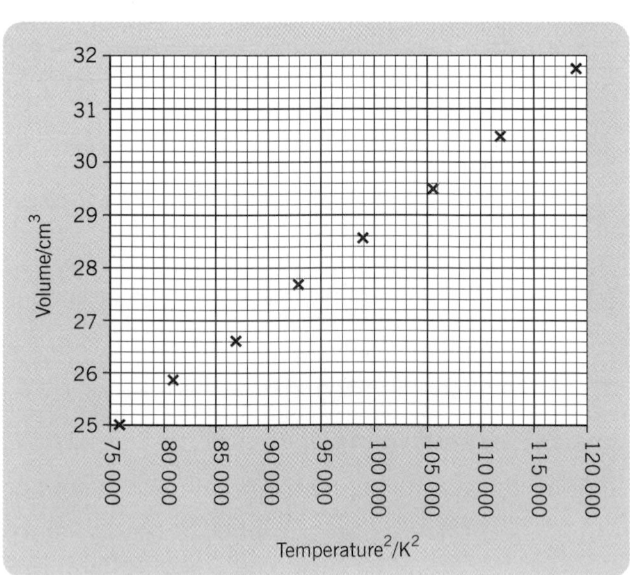

Fig. 18

4 The rate of decomposition of benzene diazonium chloride to give phenol and nitrogen was followed by measuring the volume of nitrogen collected at various times during the reaction. From this, it is possible to calculate the concentration of benzene diazonium chloride, [diazo]. The results are shown in the table.
A number of graphs can be plotted from these data; some of them are shown below (*Figs. 19–21*). Which one of them enables you to work out the relationship between concentration of benzene diazonium chloride and time? What is the relationship?

Time/s	Volume of N_2/cm^3	[Diazo]/ mol dm^{-3}
2	1.7	0.000 965
4	3.4	0.000 838
6	4.9	0.000 726
9	6.6	0.000 599
12	8.1	0.000 486
16	9.5	0.000 382
22	11.2	0.000 254
28	12.2	0.000 180

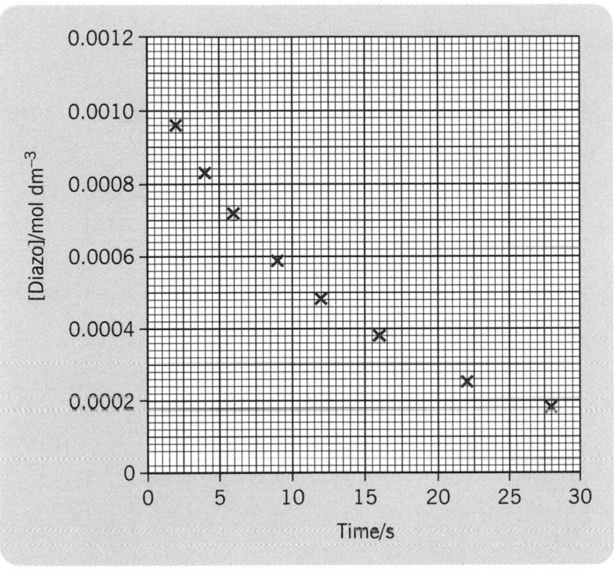

Fig. 19

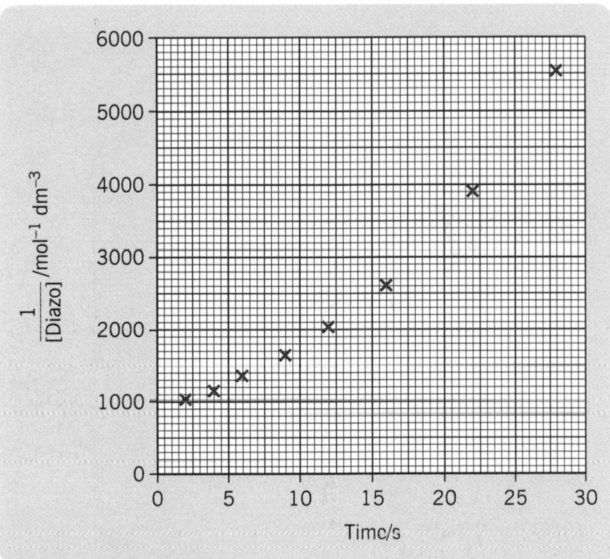

Fig. 20

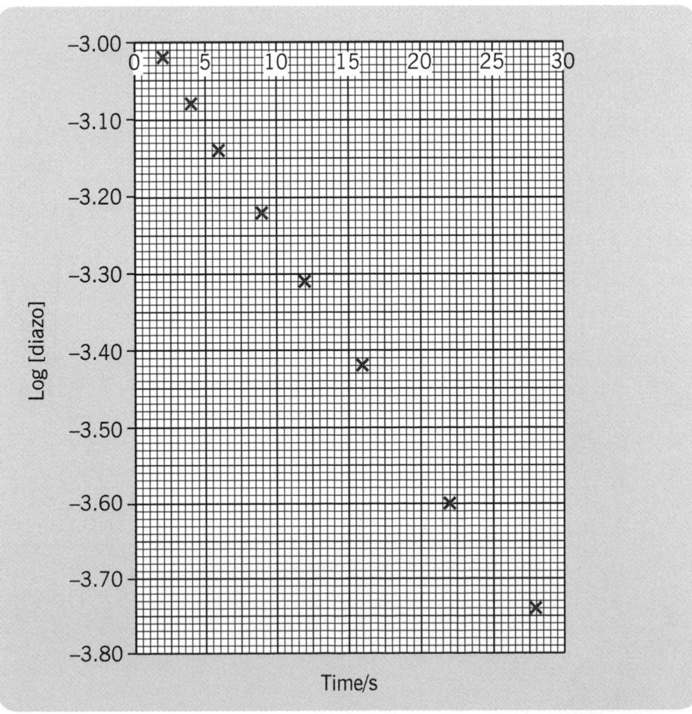

Fig. 21

5 The infra-red spectra shown below (*Figs. 22–4*) are those of propan-1-ol, propanal and propanoic acid (*but not necessarily in that order*). Use the table of data given below to match the structures to the spectra.

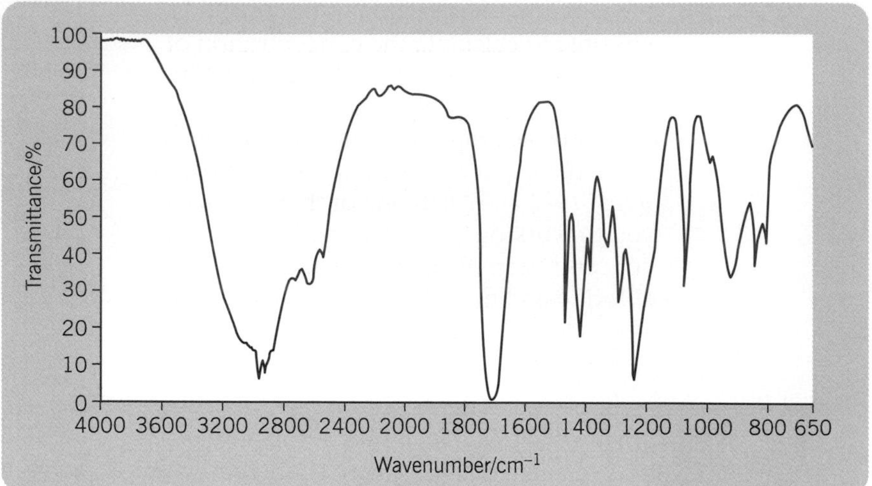

Fig. 22

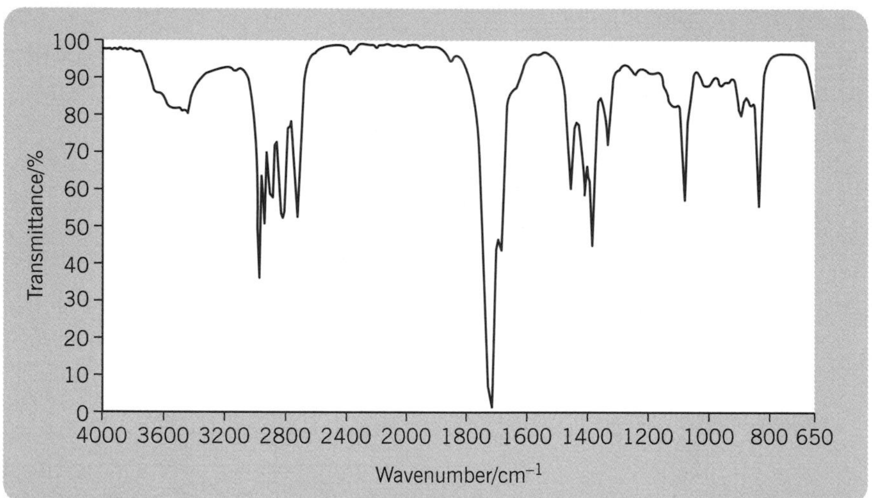

Fig. 23

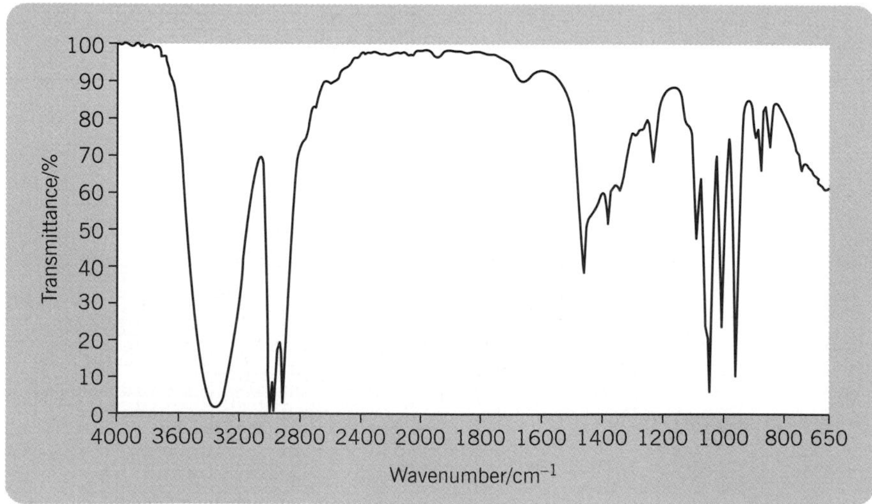

Fig. 24

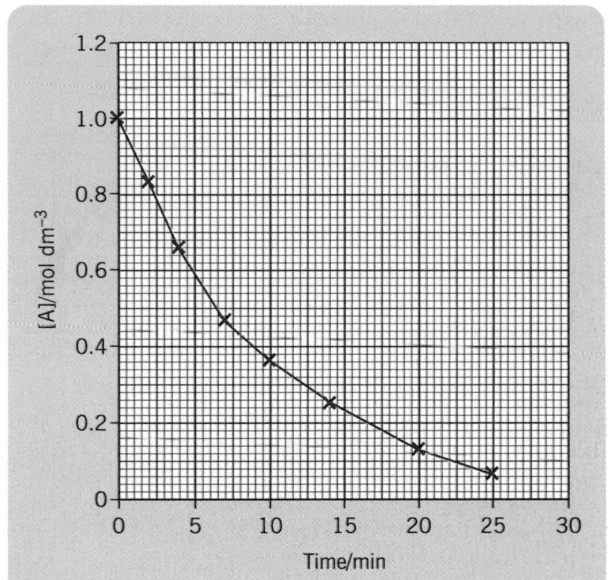

Fig. 30 Reaction of A.

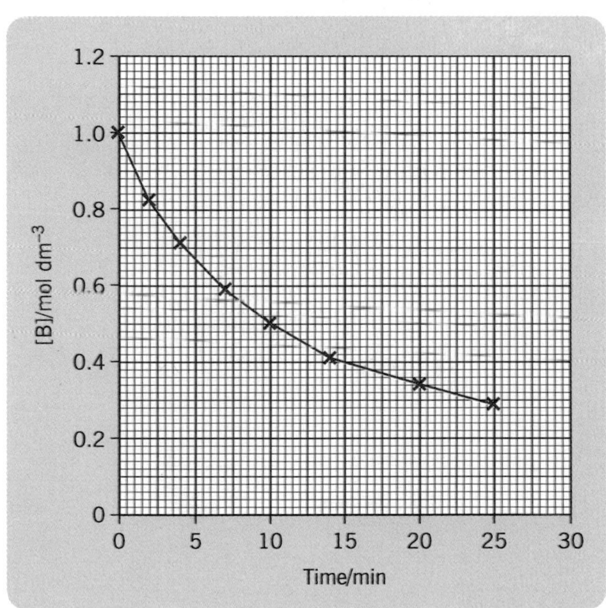

Fig. 31 Reaction of B.

concentration to fall:

(i) from 1.0 mol dm^{-3} to 0.50 mol dm^{-3}

(ii) from 0.80 mol dm^{-3} to 0.40 mol dm^{-3}

(iii) from 0.60 mol dm^{-3} to 0.30 mol dm^{-3}.

From these determine which of these reactions is first order.

(c) By drawing tangents to each curve when the concentrations are 1.0 mol dm^{-3} and 0.50 mol dm^{-3}, determine the rate of each reaction at these concentrations.

(d) Use your results from part (c) to find the order of each reaction. Do your results confirm what you found in part (b)?

10 A colorimeter was calibrated by measuring the percentage transmission of a series of solutions of iodine of known concentration. The measurements are shown in the table below.

$[I_2] \times 10^{-3}$/mol dm^{-3}	Transmission/%
0	100
1	75
2	57
3	42
4	30
5	21
6	17
7	15

Fig. 32 below is a plot of percentage transmission against concentration.

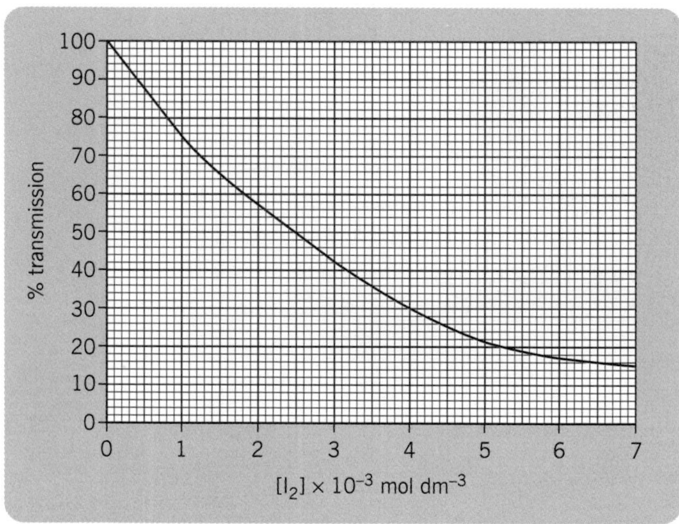

Fig. 32 *Transmission of iodine solutions.*

Use the graph to obtain:

(a) the percentage transmissions of solutions of iodine of concentrations

(i) 1.50×10^{-3} mol dm^{-3} (ii) 0.43 mol dm^{-3}

(b) the concentrations of solutions of iodine which have transmissions of

(i) 80% (ii) 20%.

11 In an experiment to measure the rate of reaction between iodine and propanone in acidic solution, various mixtures of these reagents were placed in a colorimeter calibrated as described in question 10. The time for the initial iodine concentration to fall to half its initial value was measured.

For one particular mixture, the initial colorimeter reading was 40% transmission. What would the percentage transmission be when the iodine concentration had fallen to half the initial value?

12 Mixtures were prepared containing the volumes of 1.00 mol dm^{-3} aqueous ammonia and 0.250 mol dm^{-3} copper sulphate shown in the table below. A colorimeter was used to measure the intensity of the blue colour of the copper–ammonia complex produced (as percentage absorption). These readings are shown in the table.

Vol. Cu^{2+}/cm^3	1.0	2.5	4.0	5.5	7.0	8.5
Vol. NH_3/cm^3	9.0	7.5	6.0	4.5	3.0	1.5
% absorption	9	22.5	36	40.5	27	13.5

A graph of volume of ammonia solution against absorption is shown (*Fig. 33*). Use this to deduce the volume of ammonia that will cause the maximum absorption.

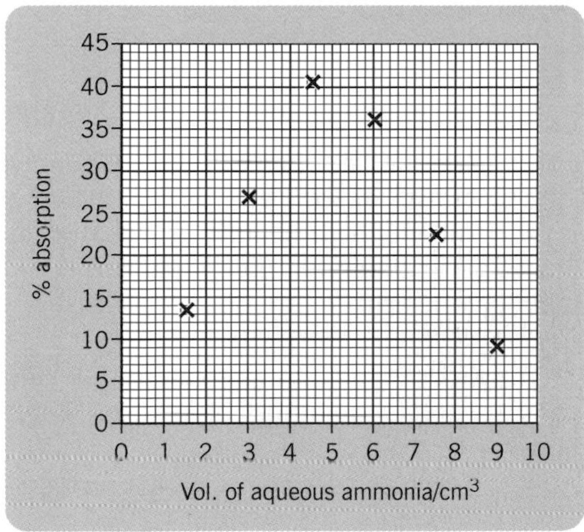

Fig. 33 *Absorption of the copper–ammonia complex.*

This corresponds to the mixture containing copper and ammonia in the exact ratio in which they react. Calculate the number of moles of ammonia and copper ions present in this solution. From this deduce the ratio in which they react, and suggest the formula of the complex produced.

13 10.0 cm^3 of 2.00 mol dm^{-3} sodium hydroxide was placed in a polystyrene cup and its temperature measured. Hydrochloric acid was added in 1.0 cm^3 portions. After each addition, the mixture was stirred and the temperature recorded. The results obtained are shown in the table.

Vol. of HCl added/cm^3	0.0	1.0	2.0	3.0	4.0	5.0	6.0	7.0
Temperature/°C	16.0	17.4	18.8	20.5	22.1	23.6	24.9	26.6

Vol. of HCl added/cm^3	8.0	9.0	10.0	11.0	12.0	13.0	14.0	15.0
Temperature/°C	27.8	29.2	28.9	28.2	27.4	26.9	26.1	25.2

A graph was plotted of temperature against volume of hydrochloric acid added (see *Fig. 34*).

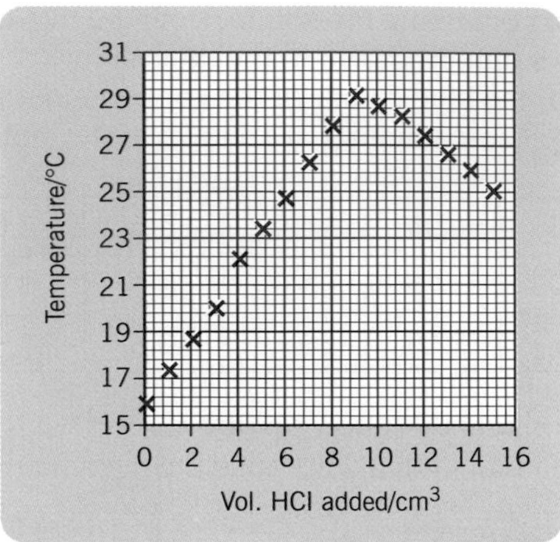

Fig. 34 *Neutralisation of sodium hydroxide.*

The temperature rises after each addition, until all the sodium hydroxide has been neutralised. Addition of further hydrochloric acid causes the temperature to fall. The end-point of the titration is the volume of hydrochloric acid that would produce the greatest temperature rise. Estimate this volume from the graph. Use it to calculate the concentration of the hydrochloric acid.

14 Thiosulphate ions, $S_2O_3^{2-}$, react with acid to give sulphur dioxide, sulphur and water. The rate of this reaction was measured at a number of different temperatures. The results are shown in the table below.

Temperature/ K	Rate × 10^{-3}/ mol dm^{-3} s^{-1}
293	7.08
303	12.3
313	21.3
323	33.4
333	55.0

It can be shown that rate (r) and temperature (T) are related by the following equation:

$$\ln r = \text{const} - \frac{E_a}{RT}$$

where E_a = activation energy of the reaction

R = gas constant

T = temperature.

The graph of ln r against $\frac{1}{T}$ is shown below (*Fig. 35*). Use the graph to obtain the value of E_a.

(Gas Constant, $R = 8.31$ J K^{-1} mol^{-1}.)

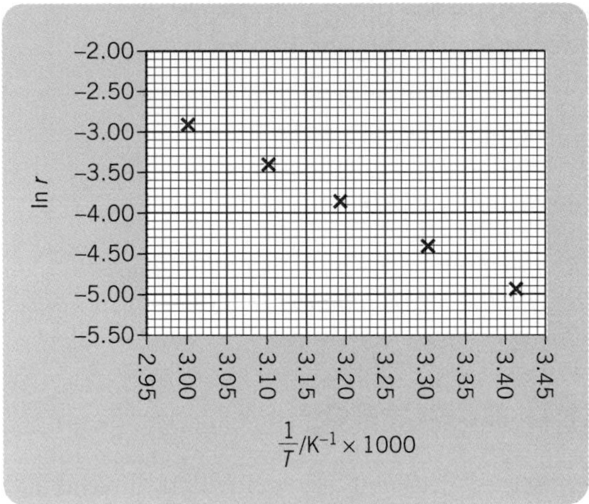

Fig. 35 *Reaction of $S_2O_3^{2-}$ with acid.*

Chapter 10

Experimental error

After completing this chapter you should:

- *understand the meaning of the term 'error'*
- *be able to estimate the error in a single reading*
- *understand the meaning of the term 'percentage error'*
- *be able to estimate the error in an experimental result based on more than one reading.*

10.1 Limited accuracy

Whenever you take any measurement, it is made with limited accuracy. If you travel from home to a town about 100 miles away and the journey takes about 3 hours, your average speed is about 33 m.p.h. If you used your calculator to work this out you would obtain the following:

$$\text{Average speed} = \frac{\text{distance}}{\text{time}} = \frac{100\ \text{(miles)}}{3\ \text{(hours)}} = 33.333\ 333\ 33\ \text{m.p.h.}$$

Of course, you would not quote the answer given above, as no one would take you seriously!

The reason for this is that you have travelled *about* 100 miles in *about* 3 hours. The most accurate figure you will have for the distance will probably be taken from the car's speedometer. That raises questions about the accuracy with which that has been manufactured and calibrated.

For most journeys, the time taken is usually based on statements such as 'we started just after 2 o'clock and we were in the pub by about 5.15'. So the time of 3 hours is very approximate. This means that the value of average speed is very approximate. 33.333 333 33 m.p.h suggests a degree of precision that is not justified. Even 33 m.p.h. may be unreliable; it might be more honest to take an average speed of between 30 and 35 m.p.h.

Results can be:

- **quantitative**, i.e. with a numerical answer, or
- **qualitative**, e.g. 'a compound contains a carbonyl group'.

Any chemistry experiment that has a quantitative result has limited precision, because each measurement is made with apparatus manufactured and calibrated to within certain limits. For example, if you have used a pipette to measure 25 cm^3 of liquid for a titration, it was probably marked 'Class B' which means that it is reliable only to ± 0.1 cm^3. Thus, your 25 cm^3 sample measured was actually 25 ± 0.1 cm^3, or between 24.9 and 25.1 cm^3. You should quote the volume as 25.0 cm^3 to reflect this.

The uncertainty in the volume delivered using the pipette will obviously affect the uncertainty of any calculation you carry out that uses that value. We need to consider how the uncertainty can be calculated. This involves consideration of **errors**.

10.2 What is 'error'?

First we need to clear up a common misunderstanding. When asked to identify sources of error in their experiment (an important step in improving the design of the experiment), students often say:

- 'I may have misread the thermometer.'
- 'There may have been calculating errors.'
- 'There may have been human error.'

We need to think what *exactly* they mean by these claims:

What they say	What they mean
'I may have misread the thermometer.'	'I may not be able to tell the difference between 17° and 18°.'
There may have been calculating errors.'	'I may have used the wrong formula, or pressed the wrong button on my calculator.'
'There may have been human error.'	'I may not be able to recognise the end point of my titration.' 'I may not have concentrated on what I was doing.'

These are errors, but they are, with a little care, avoidable. We are concerned with the errors that are *unavoidable*, because they are built into the design of the experiment or the apparatus.

KEY FACT *Error is a measure of the **uncertainty** in any measurement, or in the final answer.*

Thus in the pipette example already discussed, the error is ±0.1 cm^3.

Imagine that you have carried out an experiment to measure the concentration of a sample of hydrochloric acid. You would probably have done this by titrating it against a **standard solution** (one of accurately known concentration) of a base, such as sodium hydroxide, using an indicator such as phenolphthalein or methyl orange. One of the solutions would have been measured using a pipette, placed in a conical flask, and the other would have been added from a burette, approaching the end-point dropwise until the indicator changes colour, ideally with one drop of solution.

It is then normal practice to repeat the titration until **concordant titres** are obtained. Concordant titres lie within 0.1 cm^3 of each other. The concentration of the hydrochloric acid is then calculated using the volumes of acid and base used, and the concentration of the base.

Try It Yourself

Exercise 10.2.1

1 What are the main sources of error (uncertainty) in this experiment?

2 How big is each of the sources of error you have identified?

HINT *For example, we have already discussed the pipette, which normally has an error of ±0.1 cm^3.*

For simplicity, assume that when using 25.0 cm^3 of sodium hydroxide of concentration 0.100 mol dm^{-3}, an end-point was obtained after the addition of 25.0 cm^3 of hydrochloric acid.

The concentration of the hydrochloric acid is given by the following equation:

$$[HCl] = \frac{[NaOH] \times \text{volume of NaOH}}{\text{volume of HCl}} = \frac{0.100 \times 25.0}{25.0}$$

$$= 0.100 \text{ mol dm}^{-3}$$

The results quoted for this experiment would give a value for [HCl] of 0.100 mol dm^{-3}.

10.3 Calculating the error

One way of estimating the effect of the errors in the readings is to calculate the largest possible value for [HCl] and the smallest possible value. The largest is given by:

$$[HCl]_{max.} = \frac{[NaOH]_{max.} \times \textit{maximum} \text{ volume of NaOH}}{\textit{minimum} \text{ volume of HCl}}$$

and the smallest possible value for [HCl] is given by:

$$[HCl]_{min.} = \frac{[NaOH]_{min.} \times \textit{minimum} \text{ volume of NaOH}}{\textit{maximum} \text{ volume of HCl}}$$

The values needed for these two equations are shown in the following table:

Table 1 *Maximum and minimum experimental values*

Measurement	Value	Error	Minimum value	Maximum value
Volume of sodium hydroxide (pipette)/cm^3	25.0	±0.1	24.9	25.1
Volume of hydrochloric acid (burette)/cm^3	25.0	±0.2 see below*	24.8	25.2
Concentration of sodium hydroxide/mol dm^{-3}	0.100	±0.0005	0.0995	0.101 (to 3 s.f.)

Errors in measurements involving differences

* The volume of HCl in the table is obtained as the difference between the initial and final burette readings. Each reading has an error of ±0.1 cm^3. If the initial reading was 0.0 cm^3 and

the final reading 25.0 cm^3, the maximum difference could be 25.2 cm^3 (from the *highest* possible final reading and the *lowest* possible initial reading). The minimum difference could be 24.8 cm^3 (from the *lowest* possible final reading and the *highest* possible initial reading).

Estimating the error in a temperature rise or fall involves the same reasoning – it is based on the difference between initial and final temperature readings.

Many weighings involve *weighing by difference* – that is the difference between the mass of a container with contents and the empty container. If you use a balance with a tare facility, this reduces the error.

KEY FACT *Measurement based on the difference between two readings has* twice *the error of the individual readings.*

Substituting the values given in Table 1 into the equations gives the following:

$$[HCl]_{max.} = \frac{[NaOH]_{max.} \times \text{maximum volume of NaOH}}{\text{minimum volume of HCl}}$$

$$= \frac{0.1005 \times 25.1}{24.8}$$

$$= 0.102 \text{ mol dm}^{-3}$$

$$[HCl]_{min.} = \frac{[NaOH]_{min} \times \text{minimum volume of NaOH}}{\text{maximum volume of HCl}}$$

$$= \frac{0.0995 \times 24.9}{25.2}$$

$$= 0.0983 \text{ mol dm}^{-3}$$

You can see that, although the readings obtained in this experiment give the concentration of the hydrochloric acid as 0.100 mol dm^{-3}, the *errors* (*uncertainties*) in the readings mean that it could be *anywhere* between 0.0983 mol dm^{-3} and 0.102 mol dm^{-3}. A shorter way of conveying this uncertainty is to give the result as:

Concentration of hydrochloric acid = 0.100 ± 0.002 mol dm^{-3}.

Percentage error

The method shown above for estimating the error in an experimental result is rather tedious, especially if you do a lot of experimental work, and have to carry out the procedure every time – it may end up taking longer than the practical work! A much shorter method of estimating error is to use the concept of **percentage error**.

KEY FACT

$$\text{Percentage error} = \frac{\text{actual error in measurement} \times 100}{\text{measurement}}$$

Applying this to the titration readings we have been considering (*from Table 1, p. 98*), we obtain:

Table 2 *Percentage error.*

Measurement	Value	Error	Percentage error
Volume of sodium hydroxide (pipette)/cm^3	25.0	±0.1	$\frac{0.1 \times 100}{25.0} = 0.40\%$
Volume of hydrochloric acid (burette)/cm^3	25.0	±0.2	$\frac{0.2 \times 100}{25.0} = 0.80\%$
Concentration of sodium hydroxide/mol dm^{-3}	0.100	±0.0005	

Question: What is the percentage error in the concentration of sodium hydroxide, for the final box?

Answer: 0.50%.

To estimate the error in the concentration of hydrochloric acid, we need to consider how these three percentage errors combine.

For calculations in which the values are *multiplied* together or *divided* one into another:

KEY FACT *Percentage error in final value = sum of the percentage errors in each measurement*

In this case:

vol. NaOH vol. HCl [NaOH]

$$\text{Percentage error in [HCl]} = 0.40 + 0.80 + 0.50 = 1.70\%$$

The calculated value for the concentration of hydrochloric acid is 0.100 mol dm^{-3}, so the error is 1.70% of this value:

$$\text{Error in [HCl]} = \frac{0.100 \times 1.7}{100} = \mathbf{0.0017\ mol\ dm^{-3}}$$

Now, the result of the experiment can be written:

$$\textbf{Concentration of hydrochloric acid} = \mathbf{0.100 \pm 0.002\ mol\ dm^{-3}}$$

Notice that this result is the same as that obtained by the longer route already used! For most error estimates, the formula given in the Key Fact above will be sufficient to provide you with a reasonable estimate of the errors in any experiment you carry out.

HINT *This method of combining errors is only valid when the result is obtained by multiplying or dividing measurements. This will cover most situations in your course. For measurements involving* addition or subtraction *(volumes measured using a burette, etc.), you need to use the method already described on p. 98.*

10.4 Estimating error

For many measuring instruments, the manufacturers specify the **tolerance** (or error). Class B pipettes, as already mentioned, have a tolerance of ±0.1 cm^3 (marked as ±0.06 cm^3 on lab. pipettes). A similar figure applies to burettes.

Table 3 *Tolerance values of apparatus in the lab.*

Apparatus	Manufacturer's tolerance*
pipette (B) (25.0 cm^3)	±0.06 cm^3
burette (B) (50.0 cm^3)	±0.1 cm^3
volumetric flask (100 cm^3)	±0.1 cm^3
volumetric flask (250 cm^3)	±0.3 cm^3
volumetric flask (1000 cm^3)	±0.4 cm^3
measuring cylinder (100 cm^3)	±1.0 cm^3

*Figures taken from apparatus in the author's laboratory.

For balances, the error is usually implied by the number of figures in the reading. For example, a balance reading to 0.1 g can usually be assumed to have a tolerance of ±0.05 g. A balance reading to 0.01 g similarly has a tolerance of ±0.005 g.

For other measuring instruments, you can make a sensible estimate of the error by thinking how accurately you can read the scale (this will usually reflect the tolerance to which it is manufactured). For example, a thermometer reading from −10 °C to 110 °C usually has a scale with intervals of 1°. It is possible to divide these into $\frac{1}{2}$ intervals 'by eye'. Thus any reading should be accurate to within $\frac{1}{4}$°. However, thermometers are frequently used to measure temperature *difference*, so that the error of ± $\frac{1}{4}$° applies to both initial *and* final readings.

Try It Yourself

Exercise 10.4.1

1 To prepare a solution of sodium hydroxide of concentration 0.100 mol dm^{-3}, 4.00 g of sodium hydroxide needs to be dissolved to make 1 dm^3 of solution. This is done in a graduated flask which has a mark on the neck to indicate when the **contents** have a volume of 1.00 dm^3.

(a) Estimate the percentage error in weighing 4.00 g on a chemical balance.
(b) Estimate the percentage error in making 1.00 dm^3 of solution.
(c) Identify and estimate any other errors in this procedure.
(d) Estimate the percentage error in the concentration of the sodium hydroxide. Express this also as a concentration of 0.100 ± x mol dm^{-3}, where x is the actual error.

2 Estimate the error, and hence percentage error, in measuring a temperature rise of 25 °C with a −10 to 110 °C thermometer.

3 In an experiment to measure the enthalpy of combustion of methanol, a spirit burner was weighed, lit and placed beneath a metal container holding 200 cm^3 of water. The water was measured using a 100 cm^3 measuring cylinder. After the temperature of the water had risen by 20 °C, the flame on the burner was extinguished. The burner was then reweighed. Some typical results are shown below.

Mass of spirit burner/g	Initial 123.45
	Final 122.66
Temperature of water/°C	Initial 18.0
	Final 36.5

The heat gained by the water is given by the equation.

$$Q = m \times 4.2 \times T$$

where m is the mass of water in g, T the temperature rise in K, and Q the heat gained in J.

(a) Calculate the heat gained by the water.
(b) Calculate the number of moles of methanol burned.
(c) Calculate the heat released by burning 1 mol of methanol, i.e. the enthalpy change of combustion.
(d) Estimate the percentage error in each measurement in the experiment.
(e) Use your answer to part d to estimate the percentage error in the enthalpy of combustion of methanol.
(f) Suggest other sources of error not considered in your calculation, even if they cannot be quantified.

Error and graphs

There are other methods of estimating errors. One which may be appropriate to work applies to data processing involving graphs.

You have already seen in Chapter 9 that experimental results often lead to graphs that are not *perfect* straight lines or *perfectly smooth* curves. You have to draw the best straight line, or the 'line of best fit', or you have to draw a smooth curve that passes as close as possible to your experimental line.

After you have drawn the 'line of best fit', you may have to measure the gradient. If there is some uncertainty as to which is the best straight line, you can estimate the error in the gradient by considering other lines you could *reasonably* have drawn. From this you can estimate the *maximum* possible gradient and the *minimum* possible gradient. You then have an estimate of the error (uncertainty) in your gradient.

Summary

- All measurements have a degree of error (or uncertainty).
- The error in a single measurement is usually caused by the limit to the accuracy with which the measuring instrument is manufactured and calibrated.
- Manufacturers indicate the error of standard apparatus (e.g. pipettes).

- The *percentage error* in a measurement is the $\frac{\text{error} \times 100}{\text{measured value}}$.
- The *error in a calculated result* is usually the sum of the percentage errors in each measurement.
- Where a value is calculated from the *difference of two readings* (e.g. a temperature rise), the error in that value will be the sum of the errors in each separate reading.

Answers to Try It Yourself Questions

Exercise 10.2.1, p. *98*

1 There will be errors in identifying the end-point; in measuring the titre using the burette; in the concentration of the standard solution.

2 Reasonable estimates are:
±0.05 cm^3 for the end-point, assuming two concordant results
±0.2 cm^3 for the burette
±0.0005 mol dm^{-3} for the standard solution.

Exercise 10.4.1, p. *101*

1

	Measurement	Error	Reading	% error
a	balance	± 0.005 g	4.00 g	**0.125**
b	volumetric flask	± 1.0 cm^3	1000 cm^3	**100**

c One other possible source of error is the purity of the sodium hydroxide. An old sample could have absorbed water from the atmosphere; it could also have reacted with carbon dioxide from the atmosphere to form sodium carbonate. Both of these possibilities can be eliminated almost completely by using a freshly opened bottle of solid sodium hydroxide – i.e. the error from this source can be made negligible. Another source of error is insufficient mixing of the sodium hydroxide solution after it has been made up to a total volume of 1 dm^3. This results in a layer of more concentrated sodium hydroxide at the bottom of the flask. However, this too can be made negligible by inverting the flask ten or so times.

(d) 0.125 + 0.100 = 0.225%, i.e.
0.100 ± 0.0002 mol dm^{-3}.

2 There will be an error of ±0.25 °C in *each* reading, so that the error in the temperature rise (the difference of *two* readings) will be ±0.5 °C. For a 25 °C temperature rise, this gives a percentage error of 2%.

3 (a) 200 × 4.2 × 18.5 = 15 540 J

(b) Mass of methanol burned = 0.79 g
$= \frac{0.79}{32} = 0.0247$ mol

(c) $\frac{15\,540}{0.0247} = 629\,000$ J = 629 kJ (3 s.f.)

(d)

Measurement	Error	Reading	% error
Mass of water (measuring cylinder)	±1 cm^3 (1 g)	200 cm^3 (200 g)	0.5%
Temperature rise	+0.5 °C	18.5 °C	2.7%
Mass of methanol	±0.01 g	0.79 g	1.3%
Total			4.5%

(e) 4.5%

(f) Some of the heat released by the methanol will be lost. Some heat may be lost from the container of water. Some methanol may evaporate from the spirit burner.

Chapter 11

Symmetry

The other chapters in this book have been concerned with chemical calculations, *that is, problems involving numerical data requiring a numerical answer. This final chapter is concerned with a mathematical aspect of chemistry which involves* geometry.

After completing this chapter you should be able to:

- *explain what is meant by 'symmetry'*
- *understand what is meant by 'a mirror plane of symmetry'*
- *understand what is meant by 'an axis of symmetry'*
- *recognise the presence of planes and axes of symmetry in a structure*
- *describe some effects of the presence or absence of symmetry in a structure.*

11.1 Planes of symmetry

Fig. 1 shows a square. If a mirror were placed along the line XY, what you would see with the mirror in position would be exactly what you would see without the mirror. The reflection of the lower half of the figure would be the same as the top half. Putting it the other way round, the right half of the figure exactly mirrors the left half.

The figure is said to possess **symmetry**, and the imaginary mirror is called a **mirror plane of symmetry**.

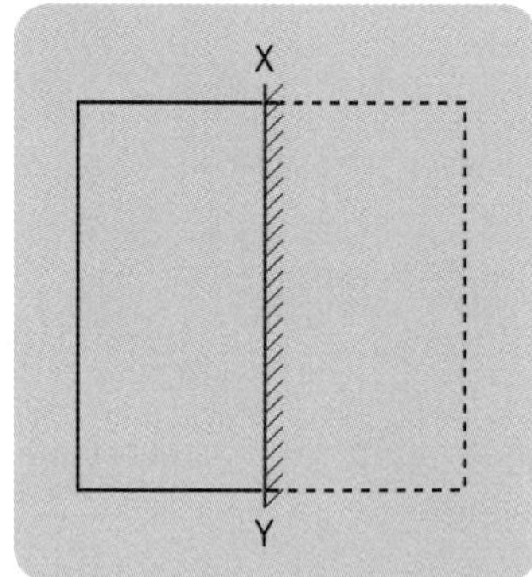

Fig. 1 *A plane of symmetry.*

11.2 Axes of symmetry

Now think of the line in *Fig. 1* as an axis. Rotate the figure through 360° about that axis. The figure comes to an identical position *twice* in that one revolution. The axis is called an **axis of symmetry**.

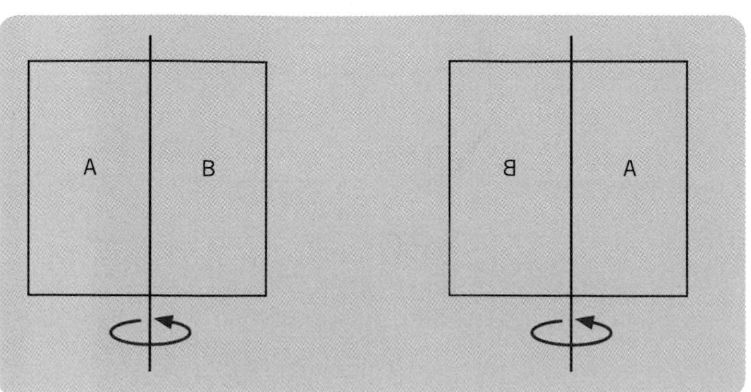

Fig. 2 *An axis of symmetry.*

Try It Yourself

Exercise 11.2.1

1 The line in *Fig. 2* above is not the only axis of symmetry of the square:

(a) How many *others* can you find?
(b) For each, how many times does the square come to an identical position in the course of one complete revolution?

2 The number of axes of symmetry is different if we replace the square with a rectangle (*Fig. 3*). How many axes of symmetry can you find for the rectangle?

Fig. 3 *A rectangle.*

3 How many axes of symmetry does the equilateral triangle (*Fig. 4*) possess?

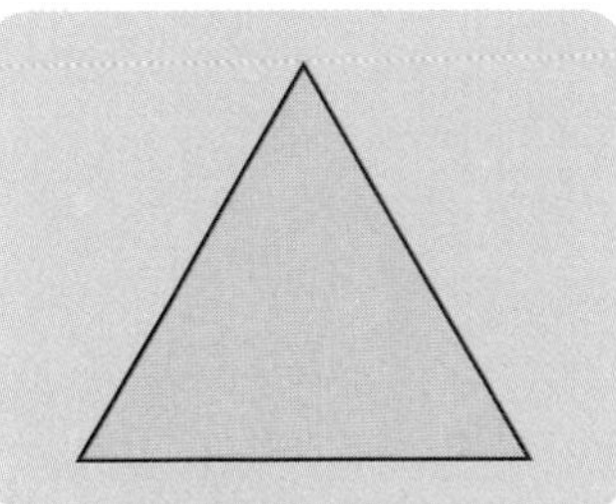

Fig. 4 *An equilateral triangle.*

4 How many axes of symmetry does the right-angled triangle (*Fig. 5*) possess?

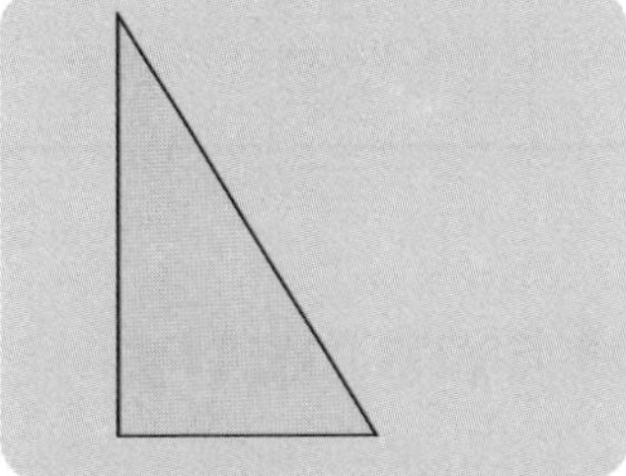

Fig. 5 *A right-angled triangle.*

11.3 Symmetry in three-dimensional shapes

The same ideas of planes and axes of symmetry can be used with three-dimensional shapes.

Try It Yourself

Exercise 11.3.1

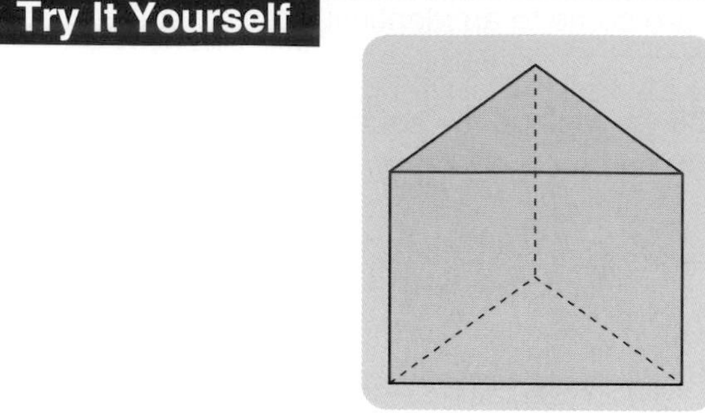

Fig. 6 *A triangular prism.*

1 *Fig. 6* shows a triangular prism. Its section is an equilateral triangle.

Besides all the symmetry of the plane triangle (*Fig. 4*) the prism has one other plane of symmetry. Where is this plane?

Example

Fig. 7 shows a cube. Through the cube is an imaginary mirror. The mirror is a plane of symmetry for the cube, because every point in front of the mirror is reflected to an exactly similar point behind it.

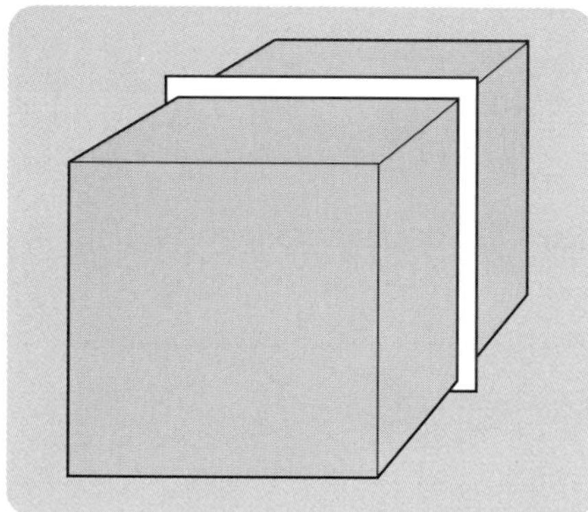

Fig. 7 *A cube showing one plane of symmetry.*

Try It Yourself

Exercise 11.3.2

1 How many planes of symmetry does the cube have in total?

HINT *If you find a plane which runs through two edges then you can say that the cube has 12 identical edges so there must be six such planes, and so on.*

11.4 Consequences of symmetry

Structures that possess symmetry (at the atomic level) have particular properties. The regularity of inorganic crystals is a consequence of their internal symmetry. The strength and malleability of metals is a consequence of the symmetry of their structure. Whilst it is unlikely that you will be examined on the symmetry of these materials, it is of interest to look at the structures to recognise this property.

Sodium chloride crystals

Fig. 8 shows part of the structure of sodium chloride crystals. *Fig. 9* shows a sodium ion surrounded by the six chloride ions that are its immediate neighbours.

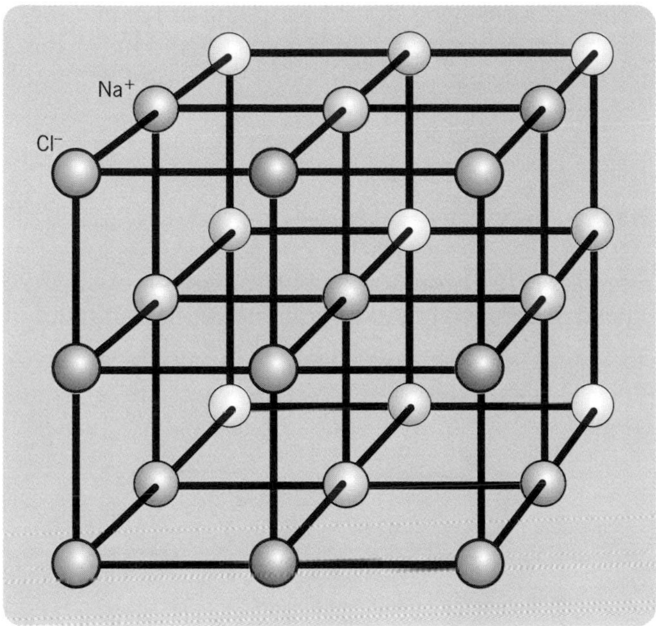

Fig. 8 *The sodium chloride structure.*

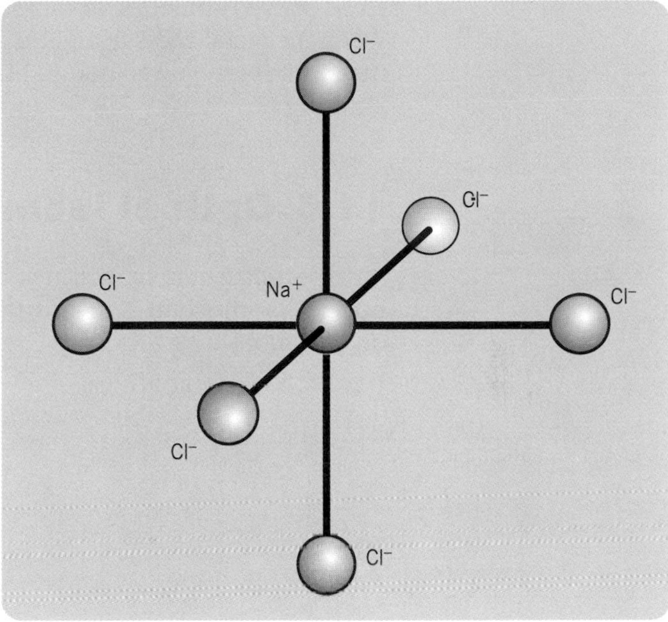

Fig. 9 *Ions in the sodium chloride structure.*

Try it Yourself

Exercise 11.4.1

1 (a) How many planes of symmetry does the structure in *Fig. 9* possess?

(b) Why do you think that sodium chloride crystals are often cubes?

Malleability of metals

The most efficient way of stacking identical spheres is to put them in sheets in which each sphere is surrounded by six others, and then stack the sheets on top of one another so that atoms of the upper sheet fit into 'dimples' between atoms in the lower sheet. This is called 'close-packing' (*Fig. 10*).

When a metal is deformed, the sheets can slip easily one over another because the metal bond is not localised between atoms. The 'sea of electrons' binds the metal ions to their near neighbours, but not to *specific* neighbours. As the sheets slip, no specific bonds are broken and the energy needed for an atom to slip from one trough to the next is only small.

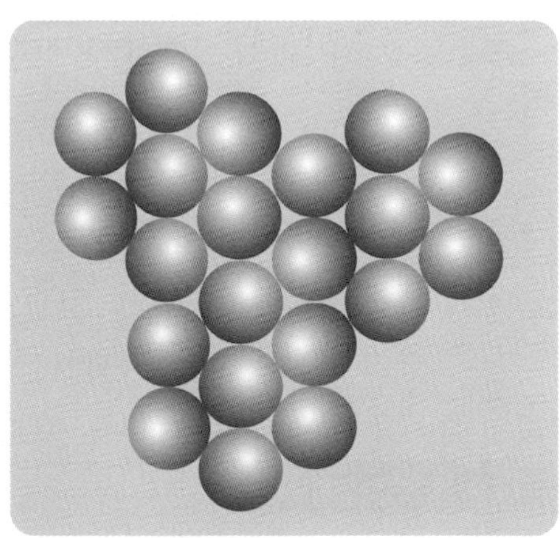

Fig. 10 *A portion of a close-packed sheet.*

In some metals, for example gold, there are four directions normal to which the atoms are close-packed. (This is called 'cubic close-packing'.) Gold is so malleable that a small gold coin could (by a skilled gold-beater) be hammered out to cover a football pitch!

Other metals (for example, iron at room temperature) have structures closely related to, but not quite, close-packed. They are malleable, but less so than gold.

If iron is heated to above 900 °C, its structure changes. The atoms re-pack into the cubic close-packed structure. In this state, the iron can easily be shaped by rolling or hammering. When the metal cools again, the change is reversed and so the iron tends to hold the shape that it has been given. Many a blacksmith has made use of this property in the past without knowing it!

11.5 Optical isomerism

An aspect of symmetry which you are more likely to encounter in your course is that of **optical isomerism**. To study this you need to appreciate the three-dimensional structure of organic molecules.

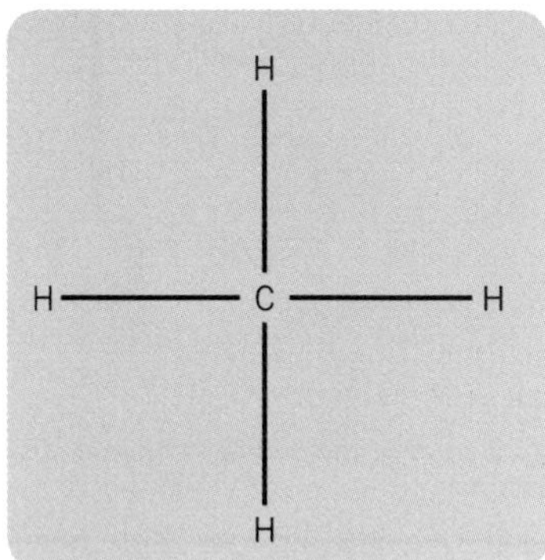

Fig. 11 *Bands in methane molecule.*

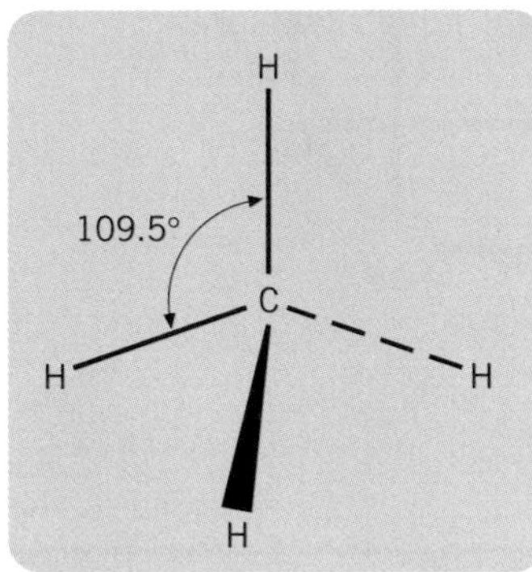

Fig. 12 *The methane molecule in three dimensions.*

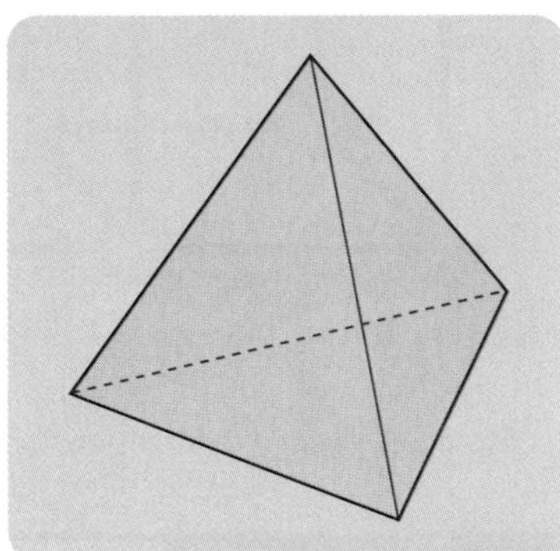

Fig. 13 *A regular tetrahedron.*

Fig. 11 shows a common way of representing the methane molecule.
Whilst this is perfectly acceptable for many purposes, the true structure of the molecule is not flat (**planar**) with bond angles of 90°, but as shown in *Fig. 12* instead.

The bond angles are 109.5°. The carbon atom is at the centre of a **regular tetrahedron** (a tetrahedron is a pyramid with a triangular base, *Fig. 13*).

The hydrogen atoms are at the corners of the tetrahedron. The bonds marked with solid lines in *Fig. 12* are in the plane of the paper but the others are not. The wedge-shaped bond comes out of the plane of the paper *towards* you, whilst the dotted bond goes into the plane of the paper *away* from you.

Try It Yourself **Exercise 11.5.1**

1 How many planes of symmetry can you find in the methane molecule?

Now replace three of the hydrogen atoms by halogen atoms: Br, Cl and I. The structure now looks like *Fig. 14*.

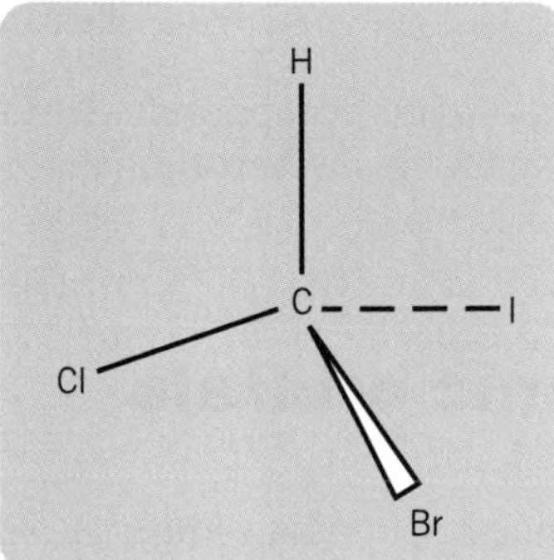

Fig. 14 *The molecule CHClBrI in three dimensions.*

Look to see if there are still planes of symmetry in the structure. There are none.

Now sketch a copy of the structure, but interchange the chlorine and bromine atoms. Better still, make Plasticine models of the two structures, or use a molecular modelling kit, if you have access to one. They are shown in *Fig. 15.* Put them side by side, turn them around as you will. You will find that:

- not only are there *no planes* of symmetry, there are *no axes* either
- one structure is a mirror image of the other
- you can never put the two in identical positions.

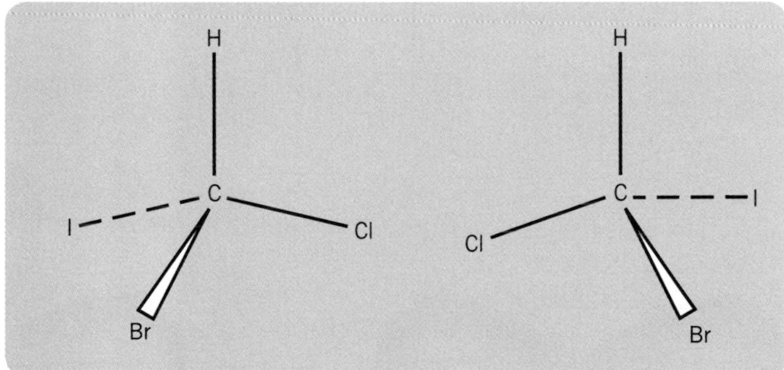

Fig. 15 *Two forms of CBrClI.*

This is true of *any* structure that has no axes and no planes of symmetry. Such a structure is said to be **asymmetrical**. The structure cannot be superposed on its mirror image. The two structures you have drawn for bromochloroiodomethane are actually different molecules; but they differ in only *one* property. They have the same melting and boiling points, solubilities, refractive indices and chemical reactions *but they rotate the plane of polarised light in opposite directions*.

KEY FACT *The absence of any plane or axis of symmetry, producing the existence of two, non-identical, mirror images, occurs in any molecule containing a carbon atom that is bonded to four different groups.*

This is very important in the field of proteins, which are made from amino acids, most of which display this property. This has implications about the *shapes* of protein molecules, but this is beyond the scope of this book.

Asymmetry of this type is not restricted to organic compounds. Certain *transition metal complexes* also exhibit the property, but again, this is beyond the scope of this book.

11.6 Symmetry in atomic orbitals

Fig. 16 shows an s atomic orbital. This shows spherical symmetry, so that bonds formed when this overlaps with an orbital from another atom can be in any direction.

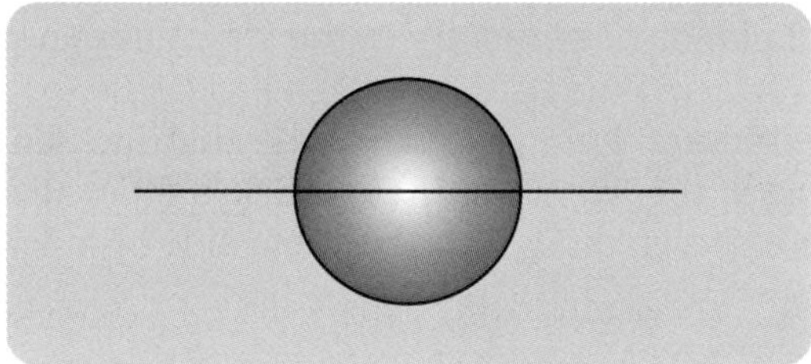

Fig. 16 *An s atomic orbital.*

Fig. 17 shows a single p orbital. There are three of these in each quantum shell, arranged at right angles to each other (p_x, p_y and p_z). They are arranged as shown in *Fig. 18*.

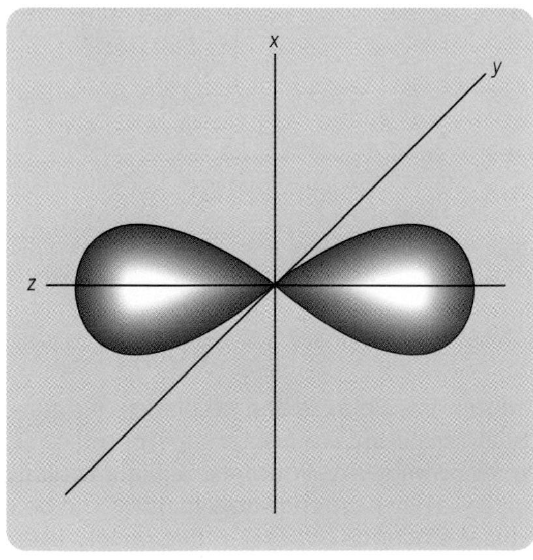

Fig. 17 *A p orbital.*

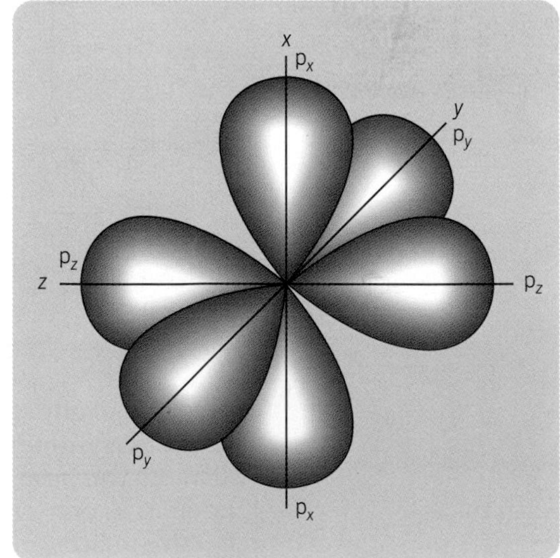

Fig. 18 *A group of p orbitals.*

One important consequence of their symmetry is that, when orbitals overlap end-on, as shown in *Fig. 19*, the atoms can rotate freely about the bond that is formed. However, if they overlap sideways, as in *Fig. 20*, no such rotation is possible.

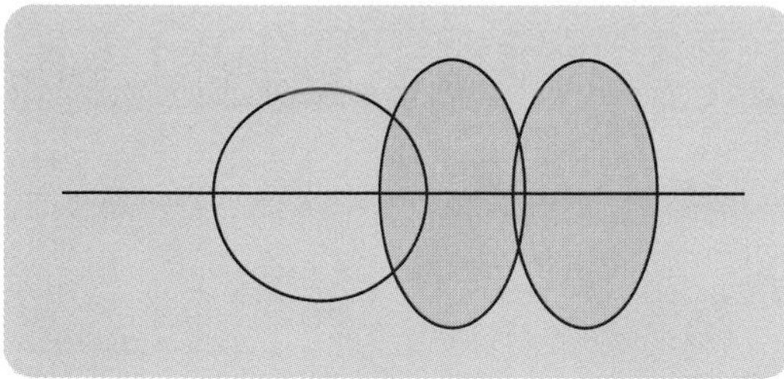

Fig. 19.

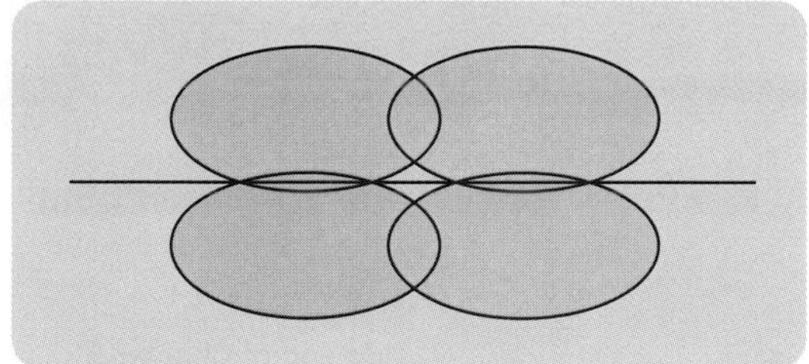

Fig. 20.

Summary

- A *plane of symmetry* divides a figure into two equal portions, each of which is a reflection of the other in that plane.
- An *axis of symmetry* is a line through a figure such that if the figure is rotated about the line through a complete revolution, it assumes an identical position more than once.
- The shape and symmetry of crystals is a result of the symmetry of the arrangement of the particles within them.
- The strength and malleability of metals are in part consequences of symmetry in their structures.
- *Optical isomers* of the same compound rotate the plane of polarisation of plane polarised light in opposite directions.
- The symmetry of atomic orbitals can affect the character of the bonds they form.

Answers to Try It Yourself Questions

Exercise 11.2.1, *p. 105*

1 **(a)** 4

(b) For three of them, twice; for the fourth, four times; the dot ● represents an axis normal to the plane of the square.

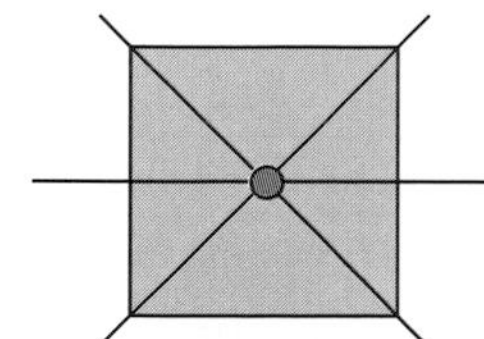

2 3

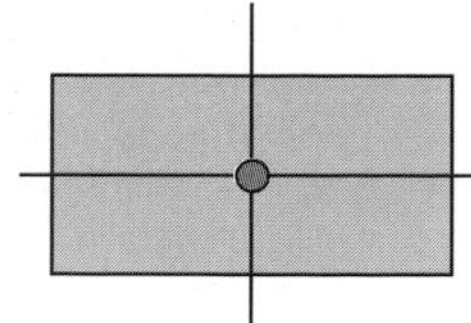

3 4

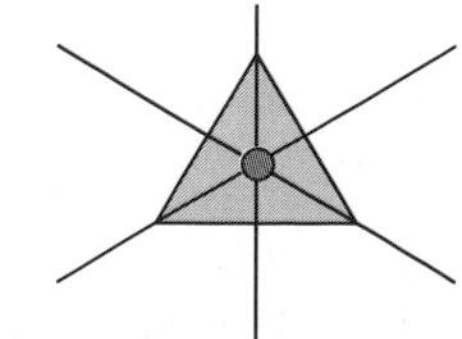

4 None

Exercise 11.3.1, *p. 106*

1 The plane bisects the length of the prism.

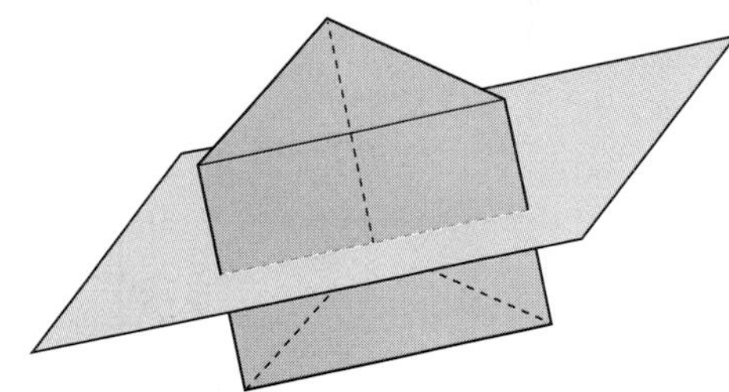

Exercise 11.3.2, *p. 106*

1 9

Exercise 11.4.1, *p. 107*

1 **(a)** 9

(b) The structure has the basic symmetry of a cube.

Exercise 11.5.1, *p. 108*

1 6 (one through each edge).

Answers to exam type questions

At end of Chapter 5

1 (a) 0.100 mol dm^{-3} (b) 0.110 mol dm^{-3}
2 (a) 1.27×10^{-3} mol (b) 38.7 cm^3
3 70.9
4 + 1 kJ mol^{-1}
5 −319.8 kJ mol^{-1}
6 −799 kJ mol^{-1}
7 −106 kJ mol^{-1}
8 (a) 1st order (b) 3.00×10^{-4} s^{-1}
9 (a) (i) and (vi) (b) (vi)
(c) 0.0100 mol^{-1} dm^3 s^{-1};
6.25×10^{-4} mol dm^{-3} s^{-1}
10 1st order with respect to ICl, 2nd order with respect to hydrogen
Rate = $k[ICl][H_2]^2$; k = 21.0 mol^{-2} dm^6 s^{-1}
11 Rate = $k[CH_3COCH_3][H^+]$; k = 1.69×10^{-3} mol^{-1} dm^3 s^{-1}
12 CH
13 Fe_2O_3
14 (a) 4.90×10^{-3} mol dm^{-3} (b) 0.67 g dm^{-3}
(c) 2.4×10^{-4}
15 15 g dm^{-3}
16 16.0 mol dm^{-3}

At end of Chapter 9

1 0.0800 mol dm^{-3}
2 (a) (i) 32% (ii) 36% (iii) 58%
(b)

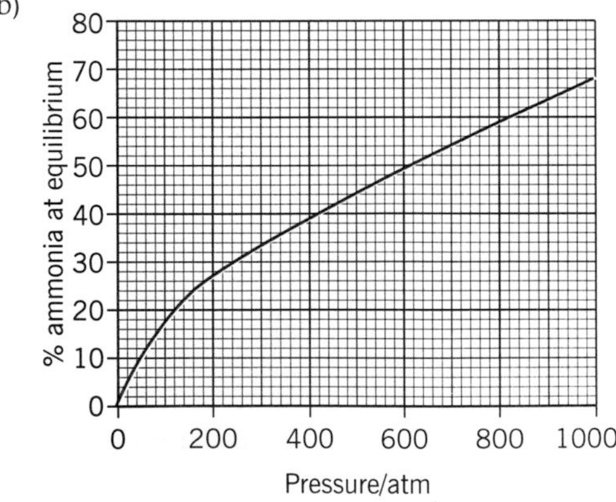

3 *Fig. 17*
4 *Fig. 21* log [diazo] = −0.028 × time − 2.95
5 *Fig. 22* = propanoic acid
Fig. 23 = propanal
Fig. 24 = propan-l-ol
6 *Fig. 25* = butanone
Fig. 26 = ethyl ethanoate
7 65.3

8

Element	Percent by mass	Percent by atoms
Carbon	19.2	9.4
Hydrogen	10.0	63.3
Nitrogen	2.8	1.4
Oxygen	64.4	25.2
Others	3.6	0.3

Although 62% of the atoms in the body are hydrogen, they are the lightest (A_r = 1). The heavier oxygen atoms (A_r = 16) account for more of the body's mass, even though there are fewer of them.

9 (a)

	Time for A/min	Time for B/min
From 1.0 mol dm^{-3} to 0.50 mol dm^{-3}	$6\frac{1}{2}$	10
From 0.80 mol dm^{-3} to 0.40 mol dm^{-3}	$6\frac{1}{2}$	$11\frac{1}{2}$
From 0.60 mol dm^{-3} to 0.30 mol dm^{-3}	$6\frac{1}{2}$	18

(b) The half-life for reaction A is approximately constant, whilst that for B depends on the initial concentration. A is thus a first order reaction.

(c)

	For reaction A	For reaction B
Rate at 1.0 mol dm^{-3}	−0.0870	−0.0952
Rate at 0.50 mol dm^{-3}	−0.0497	−0.0240

(d) Halving the concentration (approximately) halves the rate of reaction A, making it first order (confirming the result in (a). This method depends very much on how accurately you are able to draw the tangent to the graph!
For reaction B, halving the concentration reduces the rate by a factor of (approximately) 4, showing that this is a second order reaction. This is consistent with the results obtained in part (a).

10 (a) (i) 65% (ii) 86%
(b) (i) 0.80×10^{-3} mol dm (ii) 5.2×10^{-3} mol dm^{-3}
11 1.6×10^{-3}
12 $[Cu(NH_3)_4]^{2+}$
13 2.17 mol dm^{-3}
14 43 J mol^{-1}